Gynecologic and Obstetric Investigation

Founded 1895 as 'Monatsschrift für Geburtshilfe und Gynäkologie',
continued 1946–1969 as 'Gynaecologia' and 1970–1977 as 'Gynecologic Investigation'

Founders: A. Martin and M. Sänger
Former Editors: E. Anderes (1939–1952), Th. Koller (1939–1969), O. Käser (1954–1969),
R. Wenner (1959–1967), P. Bloch (1965–1969), W.L. Hermann (1970–1975),
P.J. Keller (1976–1984), G. Zador (1985–2001), J. Yankowitz (2002–2005)

KARGER

Printed in Switzerland
on acid-free and non-aging
paper (ISO 9706) by
Reinhardt Druck, Basel

Appears 6-weekly:
2 volumes per year
(8 issues)

Guidelines for Authors

Gynecologic and Obstetric Investigation

Submission

Manuscripts written in English should be submitted using the online submission website at:

www.karger.com/goi

or as e-mail attachment (the preferred word-processing package is MS-Word) to the Editorial Office:

goi@karger.ch

Profs. T.M. D'Hooghe and J. Yankowitz

S. Karger AG
Editorial Office 'Gynecologic and
Obstetric Investigation'
P.O. Box
CH–4009 Basel (Switzerland)
Tel. +41 61 306 1344
Fax +41 61 306 1434
E-Mail s.aeschbach@karger.ch

The manuscripts should be accompanied by a signed copyright transfer statement (please see submission website). Presentation of manuscripts should conform with the Uniform Requirements for Manuscripts Submitted to Biomedical Journals (see N Engl J Med 1997;336:309–315).

Conditions

All manuscripts are subject to editorial review. *Original Articles* should not exceed 4 printed pages (4,000 words or approx. 12 manuscript pages double-spaced), including tables, illustrations and references. *Articles on current investigation* are limited to 2 printed pages (2,000 words of 6 manuscript pages double-spaced). *Case Reports* are carefully selected and also limited to 2 printed pages: this includes statements concerning the problem, patient history, diagnosis, treatment, results and any other information pertinent to the case and a maximum of 5 references. Only reports which are really innovative (i.e. less than 5 case reports should have been published on the same subject) and which include a review of similar cases (if any) will be considered for publication. *Reviews* are welcomed; however, the Editor-in-Chief is happy to discuss potential articles with authors who would like to contribute. Submission of an article for publication implies the transfer of the copyright from the author to the publisher upon acceptance. Accepted papers become the permanent property of 'Gynecologic and Obstetric Investigation' and may not be reproduced by any means, in whole or in part, without the written consent of the publisher. It is the author's responsibility to obtain permission to reproduce illustrations, tables, etc. from other publications.

All manuscripts must be accompanied by a covering letter signed by all authors. Assurance should be given that the manuscript is not under simultaneous consideration by any other publication.

All manuscripts originating from non-English-speaking countries must be revised by a professional linguistic reviewer and it must be evident from the covering letter that this has been done.

Names, postal and e-mail addresses of four experts in the appropriate area of research should accompany each manuscript. Referees suggested should not be from the same institution as the author.

Good clinical practice: It has become mandatory that every trial in humans must first obtain approval from an independent Ethics Committee and formal, informed consent from the patients before they participate in a clinical study or experiment. In order to avoid unnecessary delay with the review of manuscripts, authors are asked to state, preferably in the Materials and Methods section, that approval and informed consent have been obtained. These two statements must also appear on the covering letter which accompanies every manuscript and is signed by each author.

Arrangement

All pages should be consecutively numbered beginning with the title page, then the text, acknowledgements, references and legends to figures. The text in original papers should be divided under the headings: Abstract, Introduction, Material(s) and Method(s), Results, and Discussion. The text in case reports should be divided under the headings: Introduction, Case Report(s) and Conclusion(s).

Title page: The first page of each paper should indicate the title, the authors' names, the institute where the work was conducted, and a short title for use as running head. NB: Authors wishing to preserve the phonetic meaning of diacritics (PubMed reduces diacritics to their root characters) must spell their names accordingly when submitting manuscripts (e.g. Müller should be Mueller).

Full address: The exact postal address of the corresponding author complete with postal code must be given at the bottom of the title page. Please also supply phone and fax numbers, as well as e-mail address.

Key words: For indexing purposes, a list of 3–10 key words in English is essential.

Abstract: Each paper needs an abstract of up to 200 words. It should be structured as follows:

Background/Aims:	What is the major problem that prompted the study?
Methods:	How was the study performed?
Results:	Most important findings?
Conclusion:	Most important conclusion?

Footnotes: Avoid footnotes. When essential, they are numbered consecutively and typed at the foot of the appropriate page.

Tables and illustrations: Tables and illustrations (both numbered in Arabic numerals) should be prepared on separate pages. Tables require a heading and figures a legend, also prepared on a separate page. For technical reasons, figures with a screen background should not be submitted. When possible, group several illustrations on one block for reproduction (max. size 180×223 mm) or provide crop marks. Electronically submitted b/w half-tone and color illustrations must have a final resolution of 300 dpi after scaling, line drawings one of 800–1200 dpi. Figure files must not be embedded in a document file but submitted separately (for detailed instructions, see http://www.karger.com/goi).

Letters to the Editor: This section is set aside for critical comments directed to a specific article that has been published in the journal. Letters should be brief (not exceeding 500 words), double spaced and limited to a maximum of 5 citations. The letters and replies should be prepared according to journal format. Illustrative material is only permitted with permission of the Editor-in-Chief. With your correspondence, please include your complete mailing address, telephone and fax numbers, and email addresses, if available. The Editor-in-Chief reserves the right to refuse letters, shorten letters, delete objectional comments, and make other changes to to comply with the style of the journal. Send all Letters to the Editor to the above address.

Color illustrations
Online edition: Color illustrations are reproduced free of charge. In the print version, the illustrations are reproduced in black and white. Please avoid referring to the colors in the text and figure legends.
Print edition: Up to 6 color illustrations per page can be integrated within the text at CHF 760.– per page.

References: In the text identify references by Arabic numerals [in square brackets]. Material submitted for publication but not yet accepted should be noted as 'unpublished data' and not be included in the reference list. The list of references should include only those publications which are cited in the text. Do not alphabetize; number references in the order in which they are first mentioned in the text. The surnames of the authors followed by initials should be given. There should be no punctuation other than a comma to separate the authors. Preferably, please cite all authors. Abbreviate journal names according to the Index Medicus system. (Also see International Committee of Medical Journal Editors: Uniform requirements for manuscripts submitted to biomedical journals. N Engl J Med 1997;336:309–315.)

Digital Object Identifier (DOI)
S. Karger Publishers supports DOIs as unique identifiers for articles. A DOI number will be printed on the title page of each article. DOIs can be useful in the future for identifying and citing articles published online without volume or issue information. More information can be found at www.doi.org.

Examples
(a) Papers published in periodicals: Sun J, Koto H, Chung KF: Interaction of ozone and allergen challenges on bronchial responsiveness and inflammation in sensitised guinea pigs. Int Arch Allergy Immunol 1997;112:191–195.
(b) Papers published only with DOI numbers:
Theoharides TC, Boucher W, Spear K: Serum interleukin-6 reflects disease severity and osteoporosis in mastocytosis patients. Int Arch Allergy Immunol DOI: 10.1159/000063858.
(c) Monographs: Matthews DE, Farewell VT: Using and Understanding Medical Statistics, ed 3, revised. Basel, Karger, 1996.
(d) Edited books: Parren PWHI, Burton DR: Anti-bodies against HIV-1 from phage display libraries: Mapping of an immune response and progress towards antiviral immunotherapy; in Capra JD (ed): Antibody Engineering. Chem Immunol. Basel, Karger, 1997, vol 65, pp 18–56.

Author's Choice™

With this option the author can choose to make his article freely available online against a one-time fee of CHF 2750.–. This fee is independent of any standard charges for supplementary pages, color images etc. which may apply. More information can be found at www.karger.com/authors_choice.

Page Charges

Each additional complete or partial page above 4 printed pages is charged to the author at CHF 290.–.

Proofs

PDF proofs are sent to the corresponding author and should be returned with the least possible delay.

Reprints

Order form and price list is sent with the PDF proofs. Orders submitted after the issue is printed are subject to considerably higher prices.

General Information

Gynecologic and
Obstetric Investigation

ISSN Print Edition: 0378–7346
ISSN Online Edition: 1423–002X

Journal Homepage: www.karger.com/goi

Publication Data: 'Gynecologic and Obstetric Investigation' is published 8 times a year. Volumes 65–66, each with 4 issues, appear in 2008.

Disclaimer: The statements, opinions and data contained in this publication are solely those of the individual authors and contributors and not of the publisher and the editor(s). The appearance of advertisements in the journal is not a warranty, endorsement, or approval of the products or services advertised or of their effectiveness, quality or safety. The publisher and the editor(s) disclaim responsibility for any injury to persons or property resulting from any ideas, methods, instructions or products referred to in the content or advertisements.

Subscription Rates: Subscriptions run for a full calendar year. Prices are given per year.

Personal subscription:

Print or Online	Print+Online combined
CHF 1222.–	CHF 1312.–
EUR 873.–	EUR 937.–
USD 1111.00	USD 1193.00

postage and handling (added to print and print+online)
CHF 48.– Europe, CHF 72.– Overseas
EUR 34.40
USD 66.00

Institutional subscription:

Print or Online	Print+Online combined
CHF 2444.–	CHF 2688.–
EUR 1746.–	EUR 1920.–
USD 2222.00	USD 2444.00

postage and handling (added to print and print+online)
CHF 60.– Europe, CHF 90.– Overseas
EUR 43.20
USD 82.40

Airmail surcharge: CHF 60.80 / USD 55.20
Discount subscription prices:
Society for Gynecologic Investigation

Back Volumes and Single Issues: Information on availability and prices of single print issues and print or electronic back volumes can be obtained from Customer Service at *service@karger.ch*.

Bibliographic Indices: This journal is regularly listed in bibliographic services, including *Current Contents®* and PubMed/MEDLINE.

Photocopying: This journal has been registered with the Copyright Clearance Center (CCC), as indicated by the code appearing on the first page of each article. For readers in the US, this code signals consent for copying of articles for personal or internal use, or for the personal or internal use of specific clients, provided that the stated fee is paid per copy directly to
Copyright Clearance Center Inc.
222 Rosewood Drive
Danvers, MA 01923 (USA)
A copy of the first page of the article must accompany payment. Consent does not extend to copying for general distribution, for promotion, for creating new works, or for resale. In these cases, specific written permission must be obtained from the copyright owner,
S. Karger AG, P.O. Box
CH–4009 Basel (Switzerland).

Subscription Orders:
Orders can be placed at agencies, bookstores, directly with the Publisher

S. Karger AG
Medical and Scientific Publishers
P.O. Box
CH–4009 Basel
Switzerland
(for courier services only:
Allschwilerstrasse 10
CH–4055 Basel)
Tel. +41 61 306 11 11
Fax +41 61 306 12 34
E-Mail karger@karger.ch
www.karger.com

Change of Address:
Both old and new address should be sent to the subscription source.

or further Karger offices or representatives:

France:
Librairie Medi-Sciences Sarl
36, bd de Latour-Maubourg
75007 Paris
France
Tél. +33 (0) 1 45 51 42 58
Fax +33 (0) 1 45 56 07 80
E-Mail librairie@medi-sciences.fr
www.medi-sciences.fr

Germany:
S. Karger GmbH
Postfach
79095 Freiburg
Deutschland
(Hausadresse: Lörracher Strasse 16A
79115 Freiburg)
Tel. +49 761 45 20 70
Fax +49 761 45 20 714
E-Post information@karger.de
www.karger.de

India, Bangladesh, Sri Lanka:
Panther Publishers Private Ltd.
33, First Main
Koramangala First Block
Bangalore 560 034
India
Tel. +91 80 25505 836
Tel. +91 80 25505 837
Fax +91 80 25505 981
E-Mail panther_publishers@vsnl.com
www.pantherpublishers.com

Japan:
Karger Japan, Inc.
Yushima S Bld. 3F
4-2-3, Yushima, Bunkyo-ku
Tokyo 113-0034
Japan
Tel. +81 3 3815 1800
Fax +81 3 3815 1802
E-Mail publisher@karger.jp

China, Taiwan and Malaysia:
Karger China
Suite 409, Apollo Building
1440 Central Yan An Road
Shanghai 200040
China
Tel. +86-21-6133 1861
Fax +86-21-6133 1862
E-Mail karger.ray@gmail.com

South America and Central America:
Cranbury International LLC
7 Clarendon Ave., Suite 2
Montpelier, VT 05602
USA
Tel. +1 802 223 6565
Fax +1 802 223 6824
E-Mail
eatkin@cranburyinternational.com
www.cranburyinternational.com

United Kingdom, Ireland:
S. Karger AG
c/o London Liaison Office
4 Rickett Street
London SW6 1RU
United Kingdom
Tel. +44 (0) 20 7386 0500
Fax +44 (0) 20 7610 3337
E-Mail uk@karger.ch

USA:
S. Karger Publishers, Inc.
26 West Avon Road
P.O. Box 529
Unionville, CT 06085
USA
Toll free: +1 800 828 5479
Tel. +1 860 675-7834
Fax +1 860 675-7302
E-Mail karger@snet.net

KARGER

© 2008 S. Karger AG, Basel

Fax +41 61 306 12 34
E-Mail karger@karger.ch
www.karger.com

The Journal Home Page is available at:
www.karger.com/goi

Gynecologic and
Obstetric Investigation

Novel Therapeutic Targets for GnRH Analogues in the Treatment of Endometriosis and Current Approaches to Optimizing GnRH Analogue Therapy

Proceedings of a Meeting Held during
the 10th World Congress of Endometriosis
Melbourne, March 2008

Guest Editors

Hiroshi Hoshiai, Osaka
Naoki Terakawa, Tottori

21 figures, 2 in color, and 1 table, 2008

KARGER

Basel · Freiburg · Paris · London · New York · Bangalore ·
Bangkok · Shanghai · Singapore · Tokyo · Sydney

The production of this supplement was supported by an educational grant from Takeda Pharmaceutical Company Ltd, Osaka, Japan. Editorial services for the preparation of this supplement were provided by Bluewolf Communications Ltd, Cambridgeshire, UK.

Disclosure Statement Guest Editors

Professor Hoshiai and Professor Terakawa have no conflict of interest to disclose.

S. Karger
Medical and Scientific Publishers
Basel · Freiburg · Paris · London
New York · Bangalore · Bangkok
Shanghai · Singapore · Tokyo · Sydney

Disclaimer
The statements, opinions and data contained in this publication are solely those of the individual authors and contributors and not of the publisher and the editor(s). The appearance of advertisements in the journal is not a warranty, endorsement, or approval of the products or services advertised or of their effectiveness, quality or safety. The publisher and the editor(s) disclaim responsibility for any injury to persons or property resulting from any ideas, methods, instructions or products referred to in the content or advertisements.

Drug Dosage
The authors and the publisher have exerted every effort to ensure that drug selection and dosage set forth in this text are in accord with current recommendations and practice at the time of publication. However, in view of ongoing research, changes in government regulations, and the constant flow of information relating to drug therapy and drug reactions, the reader is urged to check the package insert for each drug for any change in indications and dosage and for added warnings and precautions. This is particularly important when the recommended agent is a new and/or infrequently employed drug.

KARGER

Fax +41 61 306 12 34
E-Mail karger@karger.ch
www.karger.com

Vol. 66, Suppl. 1, 2008

Gynecologic and
Obstetric Investigation

Contents

Gynecol Obstet Invest 2008;66(suppl 1):1–2
DOI: 10.1159/000148024

Published online: October 21, 2008

Introduction

Hiroshi Hoshiai, Kinki University
Naoki Terakawa, Tottori University

This publication summarizes the presentations and discussions that took place at a round-table meeting that was held to review the current and future role of gonadotropin-releasing hormone (GnRH) analogues in the management of endometriosis. The meeting took place on 12 March 2008 during the 10th World Congress on Endometriosis in Melbourne, Australia. The program for the meeting comprised state-of-the-art scientific and clinical presentations given by an international panel of experts.

In the first session, presentations by Dr. Osuga (Japan) and Dr. Tesone (Argentina) gave a pathophysiological perspective on novel therapeutic targets in endometriosis and the role and underlying mechanism of action of GnRH analogues in this disease based on data from the latest preclinical studies. In the second session, Dr. Laufer (USA) gave an expert perspective on optimizing the treatment of endometriosis in an important patient group: adolescents.

It has previously been considered that the therapeutic effect of GnRH analogues on endometriosis can be explained by the suppression of estrogen levels due to pituitary downregulation; however, Dr. Osuga described recent exciting research by his group on the pathogenesis of endometriosis that suggests that the effect may be exerted by multiple mechanisms. One example is the inhibition of ovulation caused by GnRH analogues, which results in a reduction in the exposure of endometriotic lesions to midkine – a growth factor present in ovarian follicular fluid that has angiogenic, mitogenic, and chemotactic activities, and is thought to be involved in the proliferation of endometrial cells and development of endometriosis. He also described studies suggesting that the inhibition of bleeding caused by GnRH analogue therapy can reduce the exposure of endometriotic lesions to thrombin. Thrombin, which is produced in the process of coagulation, and its specific receptor, protease-activated receptor 1 (PAR1), are important factors in inflammation and cell proliferation and are thought to be involved in the pathophysiology of endometriosis. Abnormal uterine contractions have been observed in women with endometriosis, and it has been postulated that the resulting mechanical stretch might stimulate the production of pro-inflammatory mediators, such as interleukin-8 (IL-8), as has been observed in studies with endometrial stromal cell cultures. Thus, Dr. Osuga concluded, the inhibition of uterine contractions by GnRH analogue therapy, in particular during menstruation, would also block the mechanical stress on the endometrium, suppress the production of IL-8, and ultimately inhibit the development of endometriosis. He also noted the importance of the T-cell immune response in endometriosis. It has been argued that endometriosis is an inflammatory disease and its development is associated with an abnormal T-cell function. Dr. Osuga proposed a 'T-cell immune network hypothesis' to explain the etiology of the disease. Various types of T-helper (Th) cells – Th1, Th2, Th17 and regulatory T cells – each producing their own specific cytokine, interact with each other in a network. If the usual balance is disturbed, an excess of pro-inflammatory cytokines, such as Il-4 or IL-17, can lead to the development of endometriosis.

Dr. Tesone gave an overview of some of the other pathophysiological mechanisms involved in the development of endometriosis, namely apoptosis, cell growth and angiogenesis. She postulated that GnRH analogues acted directly on endometrial cells and inhibited their growth and proliferation by regulation of apoptotic and angiogenic mechanisms. Dr. Tesone's research group has performed studies which show that eutopic endometrial cells from patients with endometriosis show an increased proliferation rate and are less susceptible to cell death by apoptosis than those from subjects without the disease. Notably, the GnRH analogue, leuprorelin, was found to inhibit cell proliferation and increases the apoptotic rate in eutopic endometrial cell cultures. She suggested that this increase in apoptosis appeared to be mediated by an increase in the expression of the pro-apoptotic proteins Bax and FasL and a decrease in the expression of the anti-apoptotic protein Bcl-2. Angiogenesis – the development of new blood vessels from pre-existing vessels – is an important stage in the development of endometrial tissue, and is regulated by vascular endothelial growth factors (VEGFs) and angiopoietins. VEGF levels are known to be elevated in peritoneal fluid from patients with endometriosis and in the endometriotic tissue itself. In addition, it has been demonstrated that the expression of VEGF is potentiated by a variety of cytokines, including IL-1β. Dr. Tesone described recent studies showing that leuprorelin reduces the production of VEGF-A and IL-1β in eutopic endometrial cell cultures, suggesting a mechanism by which it could inhibit the development of endometriosis. She concluded that GnRH analogues are effective in reducing the growth of endometrial cells not only by the classical endocrine effect but via a direct effect on the cells themselves, and that this knowledge may contribute to the development of improved therapies for this debilitating disease.

Endometriosis is known to occur in adolescents as young as eight years of age and this patient group presents particular challenges in terms of differential diagnosis, variable presentation and symptoms, and choice of treatment. In addition, long-term outcomes of therapy have not been studied in this age group. Dr. Laufer reviewed the latest advances in the diagnosis and treatment of adolescent endometriosis and described how this could be optimized, and the pitfalls to be avoided. He confirmed the importance of early diagnosis in these patients in order to decrease pain and prevent disease progression – however, whether early diagnosis and treatment preserved future fertility was currently an unanswered question. He considered that surgery in this patient group was cytoreductive rather than curative and that postoperative medical therapy should be initiated regardless of disease stage.

Treatment with a GnRH agonist is one of the therapeutic options commonly utilized for endometriosis patients, and while these agents are recognized to be extremely effective in relieving symptoms, they may not be considered for long-term use due to possible adverse effects on bone mineralization. This consideration is of particular importance in adolescents who are at a critical age for accrual of bone mineral density (BMD), but Dr. Laufer advised that any potential problems of bone loss with GnRH therapy may be able to be avoided with the use of 'add back' therapy. He reported the results from a recent clinical study by his group showing that most adolescents with endometriosis receiving a GnRH agonist plus add-back therapy with norethindrone acetate had normal BMD at the hip. Dr. Laufer found that add-back therapy appeared to be a promising adjunct to GnRH agonist therapy for the prevention of bone loss in endometriosis patients since the administration of low doses of a steroid hormone (such as estradiol, norethindrone, or medroxyprogesterone) preserves bone density without negatively impacting the primary therapeutic effect of the GnRH agonist. It could also allow longer-term therapy than with a GnRH agonist alone. He recommended, however, that BMD should continue to be carefully monitored after the initial 6–8 month period of therapy and then approximately every two years in adolescent patients (over age 16) receiving long-term GnRH agonist therapy. He also stressed the importance of calcium and vitamin D supplementation.

We would like to extend our gratitude to the international faculty who took part in this meeting for their valuable contributions and informative feedback during the discussion sessions, which allowed us to gain valuable expert insight into these issues. We would also like to acknowledge the continuing generous sponsorship of Takeda Pharmaceutical Company Ltd., which made the meeting possible.

Disclosure Statement

Professor Terakawa has no conflict of interest to disclose.

Gynecologic and
Obstetric Investigation

Gynecol Obstet Invest 2008;66(suppl 1):3–9
DOI: 10.1159/000148025

Published online: October 21, 2008

Novel Therapeutic Strategies for Endometriosis: A Pathophysiological Perspective

Yutaka Osuga

Department of Obstetrics and Gynecology, University of Tokyo, Faculty of Medicine, Tokyo, Japan

Key Words

Endometriosis · Pathophysiology · GnRH analogues · T-cell immune response

Abstract

It has previously been considered that the therapeutic effects of gonadotropin-releasing hormone (GnRH) analogues on endometriosis could be explained by the suppression of estrogen levels due to pituitary downregulation; however, recent research on the pathogenesis of endometriosis suggests that these effects may be exerted by multiple mechanisms. These include the inhibition of ovulation, which results in a reduction in the exposure of endometriotic lesions to midkine, a growth factor present in ovarian follicular fluid that is thought to be involved in the proliferation of endometriotic cells and development of endometriosis. Also the inhibition of bleeding induced by GnRH analogue therapy can reduce the exposure of endometriotic lesions to thrombin, which is produced in the process of coagulation. Thrombin and its specific receptor, protease-activated receptor 1 (PAR1), are important factors in inflammation and cell proliferation and may be involved in the pathophysiology of endometriosis. Abnormal uterine contractions have been observed in women with endometriosis, and it is thought that the resulting mechanical stretch might stimulate the production of pro-inflammatory mediators, such as interleukin-8 (IL-8), as has been observed in studies with endometrial stromal cell cultures. The inhibition of uterine contractions by GnRH analogue therapy, in particular during menstruation, would block the mechanical stress on the endometrium and ultimately inhibit the development of endometriosis. Alongside the recent revolutionary progress in T-cell immunology, it has been argued that the development of endometriosis is associated with an abnormal T-cell function, and the existence of a 'T-cell immune network' is hypothesized to explain the etiology of the disease.

Copyright © 2008 S. Karger AG, Basel

Introduction

Endometriosis, the presence of endometrium-like glands and stroma outside the uterus, is a common, poorly understood, and extremely debilitating benign gynecological condition affecting mainly women of reproductive age. The etiology and pathophysiology of endometriosis are not well understood but continue to be investigated. Currently, no cure exists for the disease and treatment is directed to medical suppression, surgical excision, and alleviation of symptoms. Optimal therapeutic strategies are still the subject of considerable research.

This paper will provide a pathophysiological perspective on novel therapeutic targets in endometriosis and the role and underlying mechanism of action of gonadotropin-releasing hormone (GnRH) analogues in this disease

Yutaka Osuga
Department of Obstetrics and Gynecology, University of Tokyo, Faculty of Medicine
7-3-1, Hongo, Bunkyo-ku
Tokyo 113-8655 (Japan)
Tel. +81 3 3815 5411, Fax +81 3 3816 2017, E-Mail yutakaos-tky@umin.ac.jp

based on data from the latest preclinical studies. It will also review recent advances in our understanding of the role of T cells in the development of endometriosis.

Mechanism of Action of GnRH Analogues

It has previously been considered that the therapeutic effect of GnRH analogues on endometriosis can be explained by the suppression of estrogen levels due to pituitary downregulation; however, recent investigations into the pathogenesis of endometriosis suggest that the effect of GnRH analogues on endometriosis is exerted by multiple mechanisms. When considering possible alternative therapeutic mechanisms of action of GnRH analogues in endometriosis, other than hypoestrogenism, it is important to note that during GnRH analogue therapy the following phenomena, generally observed during normal menstrual cycles, are not seen:
- Retrograde menstruation
- Ovulation
- Focal bleeding
- Uterine contraction

These phenomena may provide clues to the possible underlying pathophysiological effects of these agents.

Retrograde Menstruation

Retrograde menstruation describes retrograde flow of endometrial tissue through the fallopian tubes into the peritoneal cavity. A study was conducted by Bulletti and colleagues to evaluate differences between patients with and without eutopic endometrium in the recurrence of ectopic endometriotic implants [1]. Endometrial ablation was performed on 14 women out of a total of 28 who were treated laparoscopically for endometriosis and recurrence of the disease was evaluated 24 months later. It was found that no recurrence occurred in those who had undergone laparoscopy plus endometrial ablation, whereas 9 out of 14 patients exhibited recurrence in the laparoscopy-only group (p < 0.001). These findings support a role for eutopic endometrium in the recurrence of endometriosis through tubal dissemination of endometrial debris and implantation of endometrial cells into the abdomen. In addition, they suggest that the absence of menstruation, as it occurs during GnRH analogue therapy, inhibits the development of endometriosis.

Anovulation

Although retrograde menstruation is an important factor in the development of endometriosis, it is not the whole story since it is known to occur in many women, but not all of them go on to develop endometriosis. It has therefore been suggested that the eutopic endometrium and the peritoneal environment of women with endometriosis may have different pathophysiological and biochemical properties compared with those of women without endometriosis.

Preovulatory follicles have a large antrum filled with follicular fluid that contains many bioactive substances. During ovulation, follicular fluid runs out of the follicle and diffuses into the peritoneal fluid. This fluid has an important role in nurturing endometriotic lesions and contains a range of cytokines and other growth factors, including the protein midkine, which has been suggested to be involved in the development of endometriosis.

Midkine is a basic, low-molecular-weight, nonglycosylated protein that is a member of the heparin-binding growth factor family and has been identified as the product of a retinoic acid-responsive gene. Although its expression is restricted to certain tissues in the adult, it is strongly induced during oncogenesis, inflammation and tissue repair. Midkine is known to have pleiotropic effects, including the induction of cell proliferation, cell migration, angiogenesis and fibrinolysis. Interestingly, midkine is known to be present at high concentrations in follicular fluid and is suggested to play important roles in folliculogenesis [2].

A study has been undertaken to compare the concentrations of midkine in the peritoneal fluid of women without or with endometriosis and those receiving GnRH analogue treatment [3]. Using a specific enzyme immunoassay, it was observed that midkine concentrations in the peritoneal fluid of women with advanced endometriosis (stages II, III and IV; r-ASRM score ≥6) were significantly higher (median: 1.21 ng/ml) than those of women without endometriosis or with stage I endometriosis (r-ASRM score ≤5; 0.96 ng/ml, p < 0.05). In contrast, the midkine concentrations observed in the peritoneal fluid of women undergoing GnRH agonist therapy were significantly lower than those of the other groups (p < 0.001) (fig. 1).

When midkine concentrations in peritoneal fluid were compared according to the menstrual phase, those in the luteal phase were significantly higher than those in the follicular phase, irrespective of the severity of endometriosis [3] (fig. 2). One possible explanation of this is that during ovulation, follicular fluid, which we know contains high concentrations of midkine, runs out of the follicle and diffuses into the peritoneal fluid, which may partly explain the higher concentrations in peritoneal fluid during the luteal phase compared with the follicular phase.

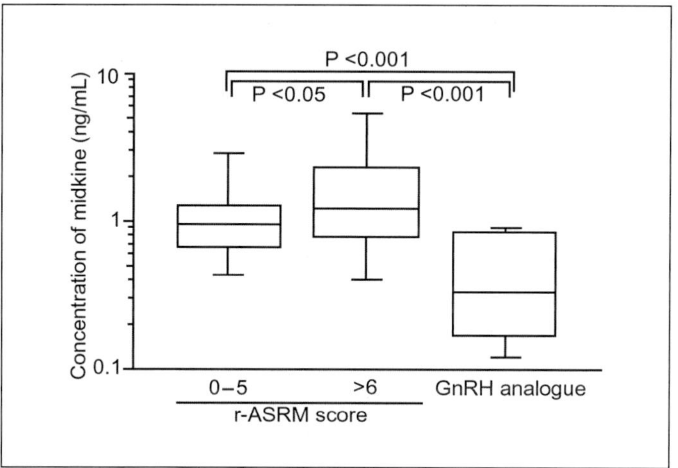

Fig. 1. Midkine concentrations in the peritoneal fluid of women without or with endometriosis and those receiving GnRH agonist treatment [3].

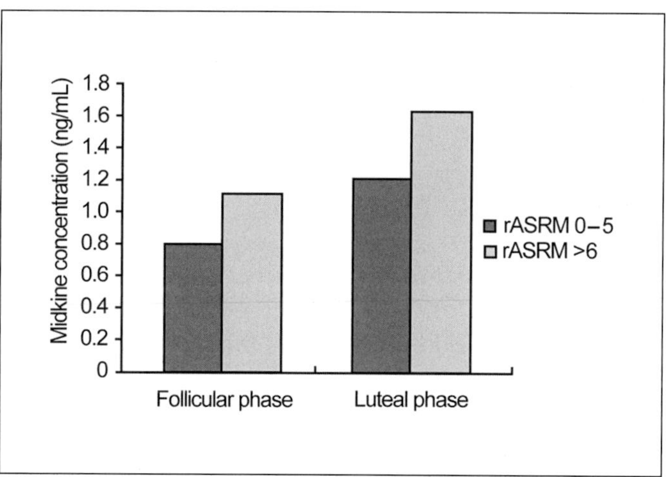

Fig. 2. Midkine concentrations in peritoneal fluid during the follicular and luteal phases of the menstrual cycle [3]. Median midkine concentration in the follicular fluid: 265 ng/ml.

The effect of midkine on the proliferation of cultured endometriotic stromal cells (ESC) has also been examined using a BrdU incorporation assay [3]. ESC were treated with midkine at different concentrations for 24 h. BrdU incorporation into DNA was significantly increased by midkine at concentrations of 100 and 1,000 ng/ml.

Overall, these results suggest that midkine may play a role in the stimulation of endometriotic cell proliferation during the development of endometriosis. In addition, it can be speculated that the inhibition of ovulation that results with GnRH analogue therapy would also block the exposure of endometriotic lesions to midkine, and subsequently suppress the proliferation of endometriotic cells and the development of endometriosis.

Absence of Focal Bleeding

Bleeding often occurs at the site of endometriotic lesions and blood clots can often be observed in endometrial cysts. Thrombin, which is produced in the process of coagulation, and its specific receptor, protease-activated receptor 1 (PAR1), are important factors in inflammation and cell proliferation and are thought to be involved in the pathophysiology of endometriosis. It has also been suggested that the inhibition of bleeding caused by GnRH analogue therapy can reduce the exposure of endometriotic lesions to thrombin and thus suppress the development of endometriosis.

PAR is a member of a group of seven transmembrane G protein-coupled receptors. As a result of the activation of PARs, proteases such as thrombin and trypsin cleave at a point within the extracellular domain and thereby unmask a new amino terminus that functions as a tethered ligand to bind back to the receptor. This property of PARs has allowed researchers to perform studies using specific agonists comprised of the amino terminal peptides to study individual PARs. To date, four PARs have been identified and characterized. PAR1, PAR3 and PAR4 are activated by thrombin; PAR2 is activated by trypsin, mast-cell tryptase, and neutrophil serine proteases. PAR1 is a typical thrombin-activating receptor and can be activated by the PAR-1 agonist peptide SFLLRN (Ser-Phe-Leu-Leu-Arg-Asp).

In view of the emerging concept of thrombin and its specific receptor, PAR1, as important players in inflammation and cell proliferation, studies have been performed to investigate whether thrombin and PAR1 might be involved in the pathophysiology of endometriosis, using a primary cell culture system of endometriotic tissues [4].

PAR1 expression was detected in cultured ESC. It was also observed that the addition of thrombin and SFLLRN to ESC cultures for 2 h significantly increased the gene expression of interleukin-8 (IL-8), monocyte chemoattractant protein-1 (MCP-1), and cyclooxygenase-2 (COX-2), important proinflammatory mediators in endometriosis. Thrombin also increased the expression of tissue factor (TF), an upstream molecule of the coagulation cascade, in ESC. Furthermore, thrombin and SFLLRN were found to stimulate the proliferation of ESC.

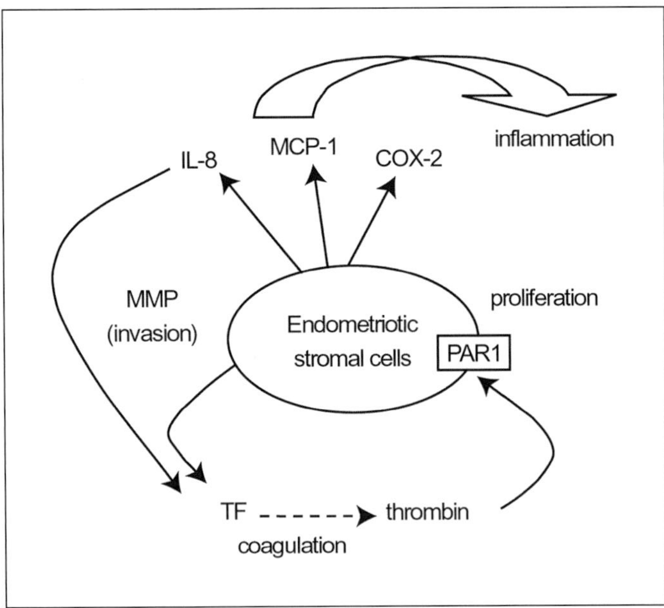

Fig. 3. The possible mechanism of action of thrombin in the development of endometriosis based on studies undertaken in endometriotic stromal cells.

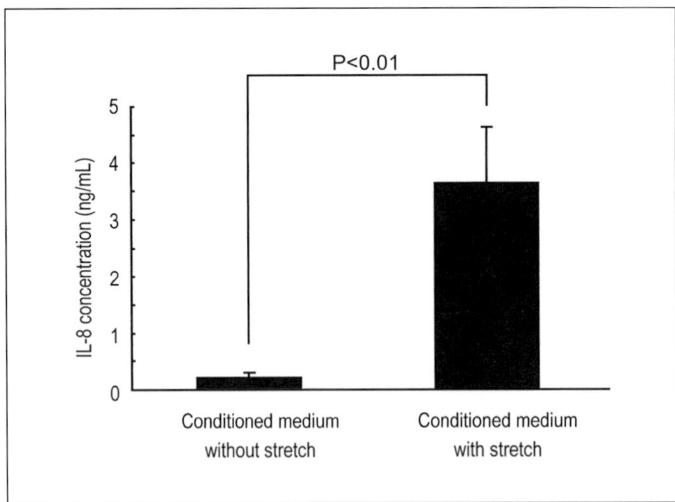

Fig. 4. Effect of conditioned medium obtained from endometrial stromal cells (EmSC) cultured with and without stretch on IL-8 secretion from EmSC [5]. Reproduced with permission from Harada et al. [5].

Matrix metalloproteinases (MMPs) are essential for extracellular matrix remodeling and may contribute to the development of endometriosis. Production of pro-MMP-9 and MMP-2 were also increased by thrombin in ESC. In this case, addition of the agonist SFLLRN did not produce a similar increase, suggesting the effect is not mediated by PAR1.

Based on the studies undertaken in ESC, the potential role of thrombin in endometriosis is summarized in figure 3. Thrombin, produced in the process of coagulation, activates PAR1 and stimulates the expression of MCP-1, IL-8, COX-2 and TF in endometriotic cells. Thrombin also stimulates the proliferation of endometriotic cells via PAR1 and increases the production of MMP via a different mechanism. The increased TF levels may further activate the coagulation cascade to produce thrombin, which subsequently enhances the production of the pro-inflammatory molecules, in a positive feedback loop, ultimately stimulating the development of endometriosis.

Thus, it is speculated that the inhibition of bleeding elicited by GnRH analogues would also block the exposure of endometriotic lesions to thrombin, and subsequently suppress the inflammation, proliferation and invasion of endometriotic cells. As a result, the development of endometriosis would be inhibited.

Absence of Uterine Contractions

Abnormal uterine contractions have been observed in women with endometriosis. It is believed that mechanical stretch induces biochemical changes in cells, and thus it has been postulated that the mechanical stretch exerted by uterine movement might stimulate the production of biochemical mediators in endometrial cells.

To investigate the effects of mechanical stretch in the endometrium, EmSC were cultured on flexible-bottomed culture plates, and cyclic stretch (25% elongation) was applied in serum-free conditions at a rate of two cycles per minute using a computer-operated cell tension system [5]. Cyclic stretch was found to increase the gene expression of IL-8 in EmSC and the secretion of IL-8 from EmSC.

Subsequent experiments also revealed some interesting characteristics. The authors investigated whether stretch-induced IL-8 expression in EmSC was regulated by autocrine/paracrine factors. They used conditioned medium obtained from EmSC cultured with cyclic stretch and added this to EmSC cultured under stationary conditions [5]. It was observed that the conditioned medium obtained from EmSC cultured with stretch induced a significant increase in IL-8 secretion and IL-8 mRNA expression from stationary-cultured EmSC, suggesting that some autocrine/paracrine mediators are involved in this phenomenon (fig. 4).

The effect of ovarian hormones on mechanical stretch-induced IL-8 production in EmSC was also investigated.

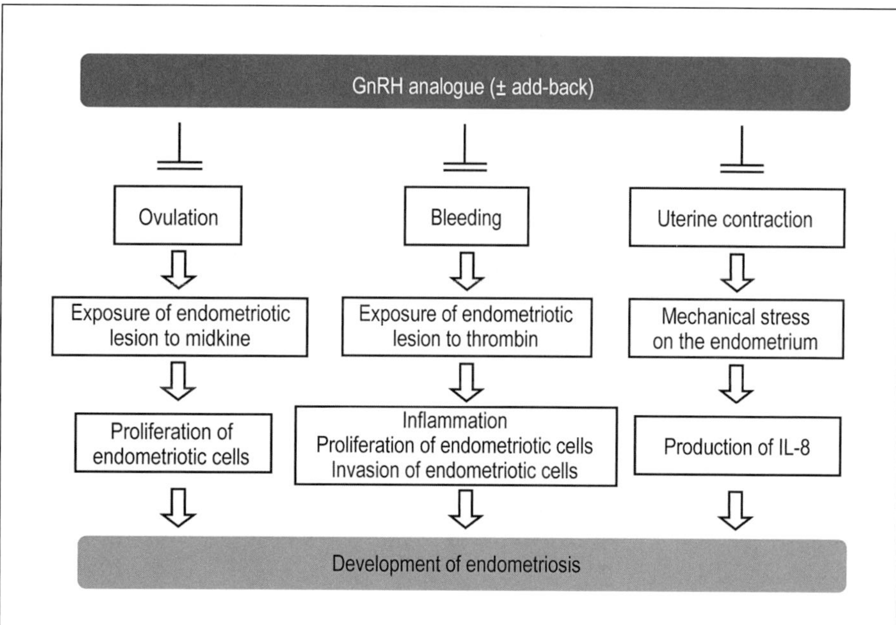

Fig. 5. Summary of the proposed effects of GnRH analogues on the inhibition of endometriosis.

The addition of progesterone, with or without estradiol, significantly suppressed stretch-induced IL-8 secretion from EmSC, whereas estradiol alone did not show any significant effects.

These findings imply that uterine movement has an impact on endometrial physiology and pathology by stimulating the production of a biochemical mediator(s) in the endometrium. It can be speculated that the inhibition of uterine contractions, particularly associated with menstruation, by GnRH analogues would also block the mechanical stress on the endometrium and subsequently suppress the production of IL-8. Ultimately, this would result in inhibition of the development of endometriosis.

In summary, based on these findings, the effect of GnRH analogues can be attributed to various events that result in the inhibition of the development of endometriosis (fig. 5). It is anticipated that these effects would be the same, regardless of whether estrogen add-back therapy was employed alongside GnRH analogue therapy.

The T-Cell Immune Response in Endometriosis

For many years it was believed that the T-helper (Th) cell family comprised only Th1 and Th2; however, recent studies have revealed that naïve Th cells are differentiated into Th1, Th2, Th17 and regulatory T cells (T-reg).

Each Th cell has been shown to secrete different cytokines; for example, Th1 secretes interferon-γ, Th2 secretes IL-4, Th17 secretes IL-17 and T-reg secretes transforming growth factor-β [6].

It has been argued that endometriosis is an inflammatory disease and its development is associated with an abnormal T-cell function [7]. Enhanced IL-4 expression has been reported in patients with endometriosis and this study clearly demonstrated that IL-4 concentration was significantly higher in ectopic endometrium compared with eutopic endometrium [8]. In view of these findings it can be hypothesized that IL-4 may have a direct effect on endometrial cells. Other studies have demonstrated alternations in the T-helper cytokine profile and in lymphocyte activation at the systemic and local levels in women with endometriosis [9]. Further investigations into the role of IL-4 have confirmed the presence of a high number of IL-4 positive cells in the endometrial tissue. In addition, IL-4 has been shown to dose-dependently stimulate the proliferation of endometrial cells [10].

A major role for the cytokine IL-17 has now been described in various models of immune-mediated tissue injury, including organ-specific autoimmunity in the brain, heart, synovium and intestines, allergic disorders of the lung and skin, and microbial infections of the intestines and the nervous system. The Th17 pathway is thought to be responsible for causing and sustaining tissue damage

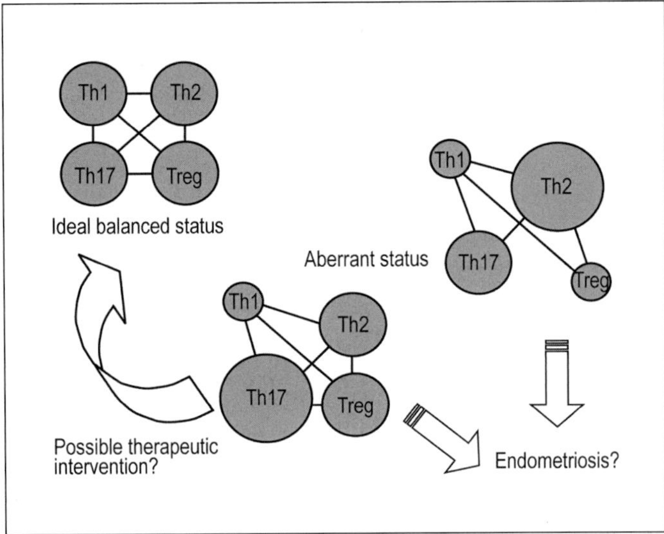

Fig. 6. The T-cell immune network hypothesis.

in these situations. It has also been shown that the Th1 pathway antagonizes the effects of the Th17 pathway.

The presence of Th17 cells has also been demonstrated in the peritoneal fluid of women with endometriosis and IL-17-positive cells have been found in endometriotic tissues [11]. To further investigate the role of IL-17 in the development of endometriosis, studies were undertaken to determine the role of IL-17 on IL-8 production, COX-2 expression, and cell proliferation of cultured ESC. IL-17 was found to enhance IL-8 secretion from ESC in a dose-dependent manner, and this was suppressed by anti-IL-17 receptor antibodies. IL-17 also enhanced the expression of COX-2 mRNA and the proliferation of ESC, as determined by increased BrdU incorporation.

Therefore, it can be concluded that IL-17 exerts proinflammatory and mitogenic effects in endometriotic tissues and may play a role in the development of endometriosis.

A recent publication by Fairbanks et al. found that patients with severe endometriosis have higher IL-12 levels irrespective of IL-18 levels in peritoneal fluid [12]. Based on these data, they claimed that, in endometriosis, an alternative pathway is involved in induction of the Th1 immune response.

Based on all these data, one hypothesis to explain the etiology of endometriosis could be the existence of a 'T-cell immune network' (fig. 6). Th1, Th2, Th17 and T-reg cells interact with each other, via cytokines or cell–cell contact, in a network. In the normal situation these pathways are balanced, but any aberrations or imbalances in these pathways or the cytokines they produce may allow endometriosis to develop. Possible future therapies might include agents targeted to restore this imbalance.

Conclusions

Data from the latest preclinical studies has shown that, in addition to their central action of suppressing estrogen levels by downregulation of the pituitary, GnRH analogues also have several other direct beneficial physiological effects on endometrial cells that result in inhibition of the development of endometriosis. Recent data have demonstrated the importance of the T-cell immune response in the etiology of this disease and may help in the development of future therapeutic agents.

Post-Presentation Discussion

Dr. Kahn, Nagasaki University, Japan: You described a positive correlation between estradiol and midkine in follicular fluid, and that midkine concentrations decrease after treatment with a GnRH analogue in peritoneal fluid of women with endometriosis. Did you also measure midkine levels in peritoneal fluid and follicular fluid after GnRH treatment with add-back therapy?

Dr. Osuga: During GnRH treatment it is very difficult to obtain follicular fluid from patients, so we only examined follicular fluid from women undergoing in vitro fertilization.

Dr. Kahn, Nagasaki University, Japan: You showed some very interesting data about the effect of mechanical stress, and the suppression of uterine contractions with GnRH analogue therapy. Estrogen has a positive correlation with levels of prostaglandin, which is known to influence a range of immune cells. Did you measure the correlation between endogenous prostaglandin levels and decreased production of IL-8 after GnRH therapy?

Dr. Osuga: We did not measure this, but it would be a very interesting experiment to do, as uterine contractions are partially under the control of estrogen.

Dr. Kitawaki, Kyoto University of Medicine, Japan: What do you propose is the mechanism of action by which GnRH analogues reduce the levels of midkine?

Dr. Osuga: During GnRH treatment ovulation does not occur. During the usual menstrual cycle at the time of ovulation follicular fluid leaks out into the peritoneal cavity and diffuses into the peritoneal fluid. Molecules, such as midkine, dissolved in the follicular fluid have an important influence on the peritoneal environment: midkine levels are 200 times higher in follicular fluid than in peritoneal fluid. During GnRH therapy, since ovulation does not occur, follicular fluid does not leak out into the peritoneal cavity, so midkine will not have its proliferative effect on endometrial tissue.

Disclosure Statement

This work was supported in part by Takeda Pharmaceutical Company Limited. No financial conflict of interest exists.

References

1 Bulletti C, DeZiegler D, Stefanetti M, Cicinelli E, Pelosi E, Flamigni C: Endometriosis: absence of recurrence in patients after endometrial ablation. Hum Reprod 2001;16:2676–2679.

2 Hirota Y, Osuga Y, Nose E, Koga K, Yoshino O, Hirata T, Yano T, Tsutsumi O, Sakuma S, Muramatsu T, Taketani Y: The presence of midkine and its possible implication in human ovarian follicles. Am J Reprod Immunol 2007;58:367–373.

3 Hirota Y, Osuga Y, Koga K, Yoshino O, Hirata T, Harada M, Morimoto C, Yano T, Tsutsumi O, Sakuma S, Muramatsu T, Taketani Y: Possible implication of midkine in the development of endometriosis. Hum Reprod 2005;20:1084–1089.

4 Hirota Y, Osuga Y, Hirata T, Yoshino O, Koga K, Harada M, Morimoto C, Nose E, Yano T, Tsutsumi O, Taketani Y: Possible involvement of thrombin/protease-activated receptor 1 system in the pathogenesis of endometriosis. J Clin Endocrinol Metab 2005;90:3673–3679.

5 Harada M, Osuga Y, Hirota Y, Koga K, Morimoto C, Hirata T, Yoshino O, Tsutsumi O, Yano T, Taketani Y: Mechanical stretch stimulates interleukin-8 production in endometrial stromal cells: possible implications in endometrium-related events. J Clin Endocrinol Metab 2005;90:1144–1148.

6 Steinman L: A brief history of T(H)17, the first major revision in the T(H)1/T(H)2 hypothesis of T cell-mediated tissue damage. Nat Med 2007;13:139–145.

7 Podgaec S, Abrao MS, Dias JA Jr, Rizzo LV, de Oliveira RM, Baracat EC: Endometriosis: an inflammatory disease with a Th2 immune response component. Hum Reprod 2007;22:1373–1379.

8 Hsu CC, Yang BC, Wu MH, Huang KE: Enhanced interleukin-4 expression in patients with endometriosis. Fertil Steril 1997;67:1059–1064.

9 Antsiferova YS, Sotnikova NY, Posiseeva LV, Shor AL: Changes in the T-helper cytokine profile and in lymphocyte activation at the systemic and local levels in women with endometriosis. Fertil Steril 2005;84:1705–1711.

10 OuYang Z, Hirota Y, Osuga Y, Hamasaki K, Hasegawa A, Tajima T, Hirata T, Koga K, Yoshino O, Harada M, Takemura Y, Nose E, Yano T, Taketani Y: Interleukin-4 stimulates proliferation of endometriotic stromal cells. Am J Pathol 2008;173:463–469.

11 Hirata T, Osuga Y, Hamasaki K, Yoshino O, Ito M, Hasegawa A, Takemura Y, Hirota Y, Nose E, Morimoto C, Harada M, Koga K, Tajima T, Saito S, Yano T, Taketani Y: Interleukin (IL)-17A stimulates IL-8 secretion, cyclooxygensase-2 expression, and cell proliferation of endometriotic stromal cells. Endocrinology 2008;149:1260–1267.

12 Fairbanks F, Abrão MS, Podgaec S, Dias JA Jr, de Oliveira RM, Rizzo LV: Interleukin-12 but not interleukin-18 is associated with severe endometriosis. Fertil Steril 2008 Feb 21 [Epub ahead of print].

Gynecol Obstet Invest 2008;66(suppl 1):10–18
DOI: 10.1159/000148026

Published online: October 21, 2008

Gynecologic and
Obstetric Investigation

The Role of GnRH Analogues in Endometriosis-Associated Apoptosis and Angiogenesis

Marta Tesone   Mariela Bilotas   Rosa Inés Barañao   Gabriela Meresman

Institute of Biology and Experimental Medicine, Buenos Aires, Argentina

Key Words

Endometriosis · Apoptosis · Angiogenesis · GnRH analogues

Abstract

It has been postulated that gonadotropin-releasing hormone (GnRH) analogues may act directly on endometrial cells and inhibit their growth and proliferation by regulation of apoptotic and angiogenic mechanisms. Eutopic endometrial cells from patients with endometriosis show an increased proliferation rate and are less susceptible to cell death by apoptosis than those from subjects without the disease. Notably, the GnRH analogue, leuprorelin, inhibits cell proliferation and increases the apoptotic rate in eutopic endometrial cell cultures, an effect that appears to be mediated by an increase in the expression of the pro-apoptotic proteins Bax and FasL and a decrease in the expression of the anti-apoptotic protein Bcl-2. Angiogenesis is an important process in the development of endometrial tissue, and it is regulated by vascular endothelial growth factors (VEGFs) and angiopoietins. VEGF levels are elevated in peritoneal fluid and endometriotic tissue from patients with endometriosis. In addition, it has been demonstrated that the expression of VEGF is potentiated by a variety of cytokines, including IL-1β. Recent studies show that leuprorelin reduces the production of VEGF-A and IL-1β in eutopic endometrial cell cultures, suggesting a mechanism by which it could inhibit the development of endometriosis. Thus, GnRH analogues appear to be effective in reducing the growth of endometrial cells, not only due to their classical pituitary endocrine effects, but also via a direct effect on the endometrial cells themselves.

Copyright © 2008 S. Karger AG, Basel

Introduction

It is generally accepted that the etiology of endometriosis (EDT) is associated with retrograde migration of eutopic endometrial tissue outside the uterine cavity. Mechanisms involved in the development of endometriotic lesions include immunological factors, apoptosis and cell growth, hormonal factors (e.g. estrogens), and angiogenesis.

Synthetic gonadotropin-releasing hormone (GnRH) analogues (both agonists and antagonists) are frequently used in the treatment of human reproductive pathologies, including EDT. The administration of long-lasting GnRH agonists reduces the endogenous secretion of gonadotropins by a mechanism of pituitary downregulation, and as a consequence they are frequently used to achieve regression of endometriotic implants due to the induction of hypoestrogenism. In addition to this central effect, it has also been demonstrated that GnRH and its agonists have a direct effect on extra-pituitary tissues,

KARGER

Fax +41 61 306 12 34
E-Mail karger@karger.ch
www.karger.com

© 2008 S. Karger AG, Basel
0378–7346/08/0665–0010$24.50/0

Accessible online at:
www.karger.com/goi

Marta Tesone
Institute of Biology and Experimental Medicine
Vuelta de Obligado 2490
1428 Buenos Aires (Argentina)
Tel. +54 11 4783 2869, ext. 218, Fax +54 11 4786 2564, E-Mail mtesone@dna.uba.ar

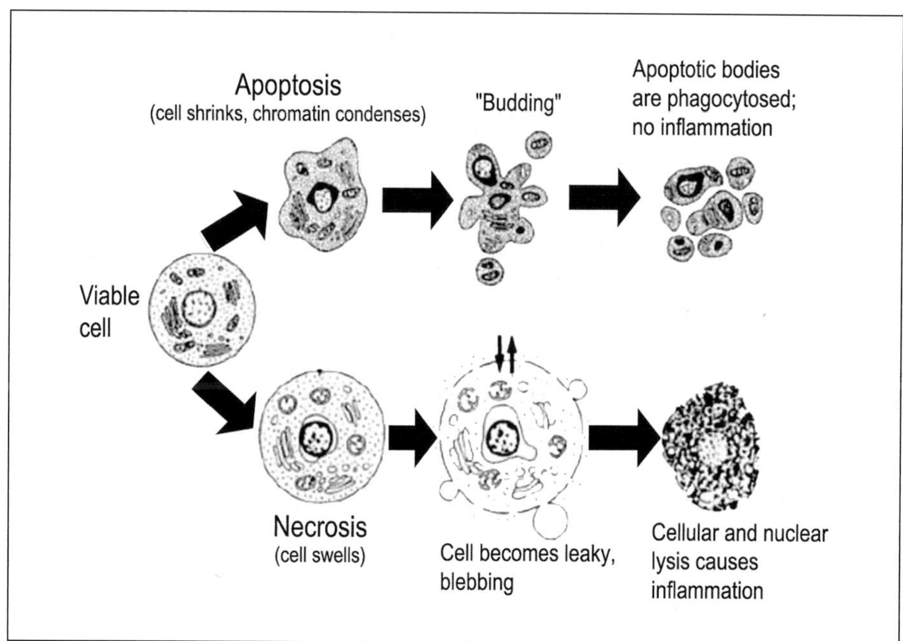

Fig. 1. Comparison of the cellular changes that occur during apoptosis and necrosis.

such as endometrium and ovary [1, 2]. There are reports demonstrating that GnRH agonists directly inhibit the proliferation of various types of steroid-dependent cancers, including those of the breast, ovary and endometrium [3–5]. In addition, a GnRH-like peptide and GnRH receptors have been isolated from both eutopic and ectopic endometrium [1].

This paper will review the underlying pathophysiological mechanisms implicated in the development of EDT and the effect of GnRH analogues in this setting.

Apoptosis

There are two different mechanisms of cell death: necrosis and apoptosis (fig. 1). During necrosis the viable cell becomes leaky, forms blebs and the resulting cellular and nuclear lysis causes a release of the cell's contents leading to inflammation. In contrast, apoptosis, or programmed cell death, also kills unwanted cells but without inducing an immune response or an inflammatory reaction. The viable cell shrinks, the chromatin condenses, budding occurs, and the apoptotic bodies are phagocytozed by white blood cells, so there is no resulting inflammation.

Apoptosis can be triggered in a cell through either the extrinsic pathway or the intrinsic pathway (fig. 2) [6]. Whichever pathway is triggered, the cell changes which

result are mediated by a family of cysteine aspartic acid-specific proteases (caspases), which occur as inactive precursors (procaspases) that are proteolytically converted to an active state following an apoptotic stimulus. The extrinsic pathway is initiated through binding of the transmembrane death receptors, such as the Fas, TNF, and TRAIL receptors, located on the cell membrane with their respective ligands. This binding triggers aggregation of the receptors and recruitment of an adaptor protein known as Fas-associated death domain protein (FADD) within the cell cytoplasm. FADD, in turn, recruits caspase-8 to form the death-inducing signal complex (DISC) which initiates the caspase activation cascade.

In contrast, the intrinsic pathway requires disruption of the mitochondrial membrane and the release of mitochondrial proteins, such as cytochrome C. Cytochrome C is released into the cytoplasm, along with the other two cytosolic protein factors, apoptotic protease activating factor-1 (Apaf-1) and procaspase-9, promoting the assembly of a caspase-activating complex termed the apoptosome, which in turn initiates the apoptotic caspase cascade. Known regulators of cytochrome C release are the Bcl-2 family of proteins, which can be either pro-apoptotic or anti-apoptotic. In addition, cross-communication occurs between the extrinsic and intrinsic pathways, and factors produced in one pathway can stimulate the other (fig. 2) [6].

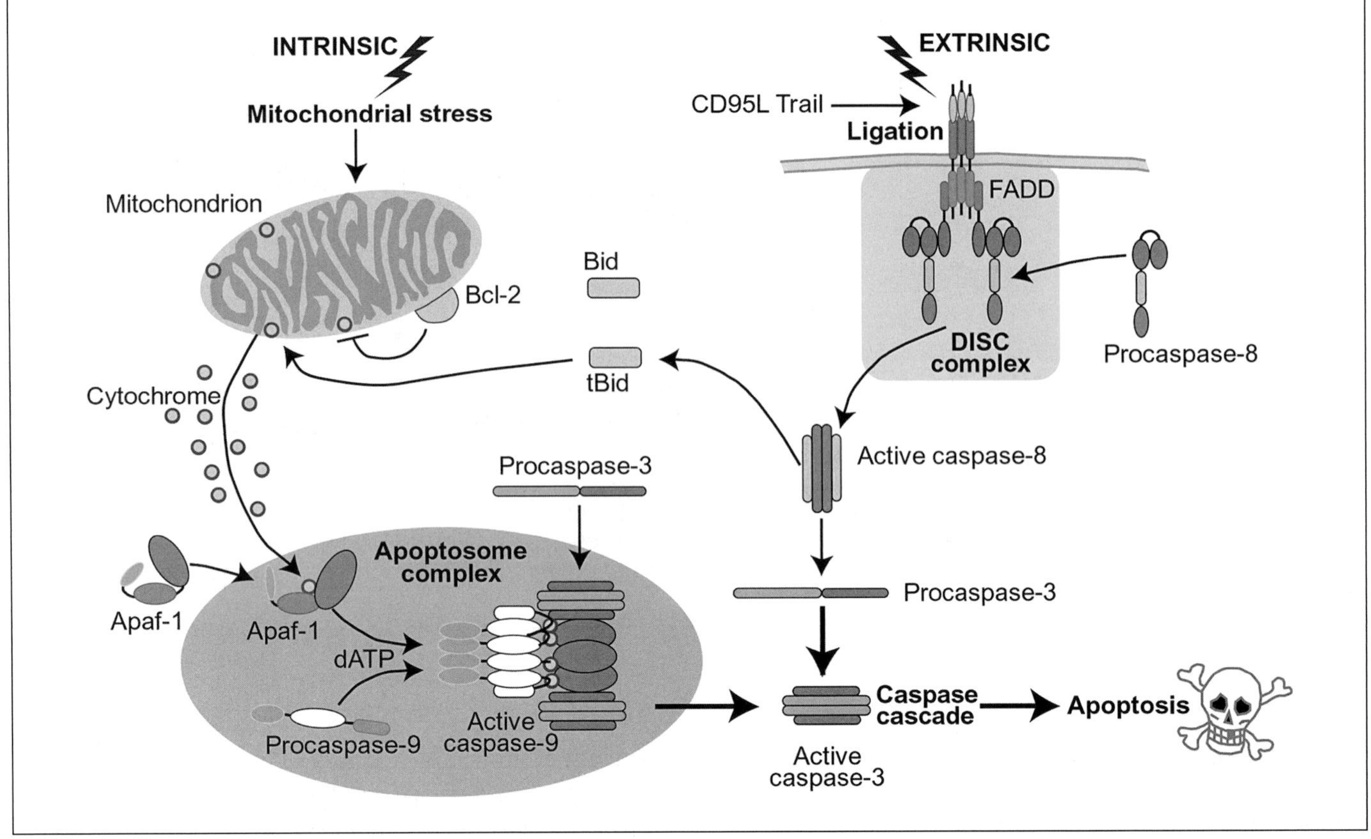

Fig. 2. The intrinsic and extrinsic apoptosis pathways. Reproduced with permission from MacFarlane and Williams [6].

Angiogenesis

There are two main processes of vascular development: vasculogenesis and angiogenesis. Vasculogenesis involves the differentiation of endothelial cells from embryologic mesoderm and the assembly of these differentiated cells into discrete blood vessels whereas angiogenesis involves the proliferation and migration of endothelial cells from pre-existing vessels and leads to the sprouting or remodeling of new vessels in a given area.

The highly regulated process of angiogenesis that occurs within the female reproductive tract is critical for normal reproduction, including follicular maturation, selection and normal function of the corpus luteum, and endometrial growth and remodeling. The pathogenesis of EDT involves the implantation of exfoliated endometrium which is dependent for its survival on the development of an extensive blood supply – by the process of angiogenesis – both within and surrounding the ectopic tissue.

There are two main regulators of angiogenesis – vascular endothelial growth factors (VEGFs) and angiopoietins. VEGF-A is a potent mitogen and migratory stimulus for endothelial cells and has been found to be involved in both physiological and pathological conditions [7]. Studies have shown that it is produced by endometrial cells and is a critical factor involved in the pathogenesis of the disease [8, 9]. VEGF levels are elevated in peritoneal fluid from patients with EDT and in the endometriotic tissue itself [8, 9]. In addition the expression of VEGF in endometrial cells has been found to be potentiated by a variety of cytokines, including IL-1 [10, 11]. Visualization of ectopic endometrial implants by laparoscopy shows them to be highly vascularized.

In view of the above evidence it has therefore been postulated that GnRH analogues interfere with endometrial cell growth by regulation of apoptotic and angiogenic mechanisms.

Gynecol Obstet Invest 2008;66(suppl 1):10–18

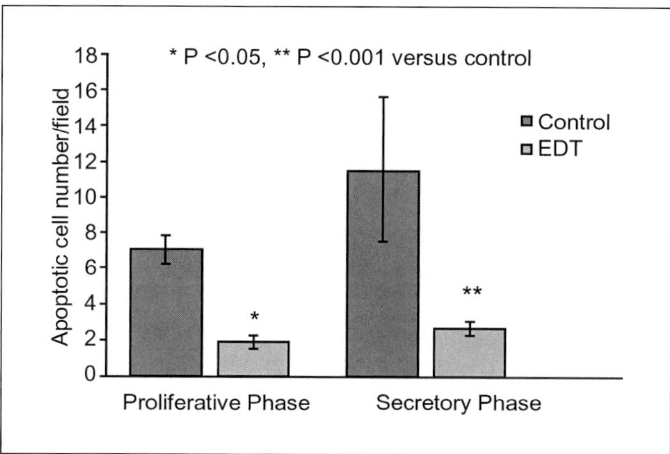

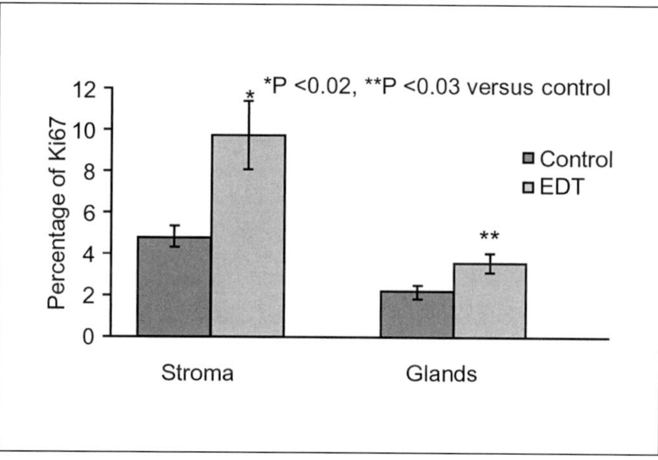

Fig. 3. Apoptosis during the proliferative and secretory phases in biopsy specimens of eutopic endometrium from control subjects and women with endometriosis (EDT) [12]. Reproduced with permission from Meresman et al. [12].

Fig. 4. Cell proliferation in stromal and glandular cells in biopsy specimens of eutopic endometrium from control subjects and women with EDT [13].

Studies of GnRH Analogues and Apoptosis

Initial studies were undertaken to evaluate apoptosis and cell proliferation in eutopic endometrium from women with and without EDT [12]. Thirty-six patients undergoing diagnostic laparoscopy for infertility participated in the study. All had regular menstrual cycles and had not received any hormonal therapy in the previous six months. Twenty-two patients had untreated EDT (stages I and II) and 14 control subjects had tubal factor infertility or unexplained infertility. Biopsy specimens of eutopic endometrium were obtained from all subjects during the proliferative phase. Apoptosis and cell proliferation were examined in these tissue sections.

The proportion of apoptotic cells in situ was determined using the TUNEL (Terminal deoxynucleotidyl Transferase Biotin-dUTP Nick End Labeling) method, in which terminal deoxynucleotidyl transferase (TdT) transfers biotin-dUTP to strand breaks of cleaved nuclear DNA that occur as a result of the apoptotic process. The biotin-labeled cleavage sites are then detected by reaction with horseradish peroxidase-conjugated streptavidin and visualized by diaminobenzidine showing brown color under the microscope. Figure 3 shows quantification of the degree of apoptosis in eutopic endometrial tissue during the proliferative and secretory phases of the menstrual cycle. In both phases, apoptosis in cells from women with EDT was significantly lower than in cells obtained from control subjects (p < 0.05 in the proliferative phase and p < 0.001 in the secretory phase vs. controls).

Cell proliferation was determined by evaluation of the nuclear protein Ki-67, which is strongly associated with cell proliferation. During interphase, it can be exclusively detected within the nucleus, whereas in mitosis most of the protein is relocated to the surface of the chromosomes. The fact that the Ki-67 protein is present during all active phases of the cell cycle, but is absent from resting cells, makes it an excellent marker for determining growth of a cell population. Ki-67 staining was observed in eutopic endometrial cells from women with EDT, suggesting that cell proliferation was occurring. This was confirmed by quantification of Ki-67, as shown in figure 4 [13]. The degree of cell proliferation (the proportion of Ki-67-positive cells) was significantly higher in both stromal and glandular cells of endometriotic tissue obtained from women with EDT than in tissue obtained from control subjects (p < 0.02 in stromal cells and p < 0.03 in glandular cells vs. controls). It can be concluded from these studies that eutopic endometrial cells from patients with EDT are less susceptible to apoptotic cell death and show an increased proliferation rate compared with cells from subjects without the disease.

As a result of these interesting findings further studies were undertaken to assess the in vitro effects of GnRH analogues on the production of factors associated with apoptosis and angiogenesis in epithelial endometrial cell cultures [14].

Biopsies from patients with EDT or control subjects were obtained as described in the earlier study. Endometrial cells were enzymatically separated and epithelial

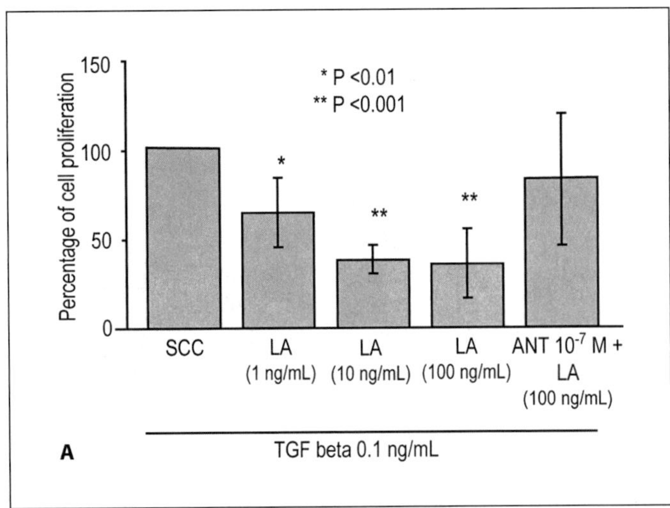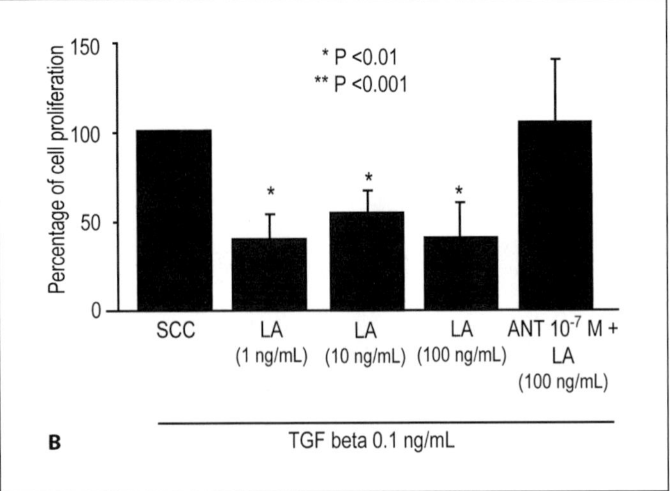

Fig. 5. Effect of leuprorelin acetate (LA) and antide (ANT) on cell proliferation in epithelial endometrial cell cultures obtained from EDT patients (**A**) or control subjects (**B**). Values are expressed as percentage of TGFβ 0.1 ng/ml stimulated cell cultures (SCC) proliferation set as 100% [15]. Reproduced with permission from Meresman et al. [15].

cells were isolated by successive centrifugation. Primary cultures were established as previously described [15]. Cells were cultured in 10% fetal bovine serum (FBS) culture medium (DMEM-F12) for 48 h and GnRH agents (a GnRH agonist: leuprorelin acetate [LA] and a GnRH antagonist: antide [ANT]) were added for 24 h to supplemented 2.5% FBS medium. Cell epithelial purity was assessed by cytokeratin immunocytochemistry (indirect immunofluorescence using mouse anti-cytokeratin 56 kDa).

Cell proliferation was measured by a thymidine incorporation method. As shown in figure 5A and B, cell proliferation in epithelial endometrial cell (EEC) cultures from both control women and EDT patients was inhibited at a range of concentrations of LA (1, 10 and 100 ng/ml), and this inhibition was reversed by the addition of ANT.

The percentage of apoptotic cells was assessed by the acridine orange–ethidium bromide technique and cells were viewed by a fluorescence microscope. When this effect was quantified (apoptotic cells counted as a percentage of the total) it was revealed that the proportion of apoptotic cells in EEC cultures from both control women and EDT patients was significantly increased in the presence of LA at concentrations of 100 ng/ml (p < 0.05 and p < 0.01, respectively) (fig. 6A, B). This apoptotic stimulation was reversed by the addition of ANT (10^{-7} M). The effect of ANT was found to be dose-dependent: at a concentration of 10^{-5} M it was found to increase the propor-

tion of apoptotic cells in EEC from patients with EDT and controls (p < 0.01), but lower concentrations were not effective (fig. 7).

These results led the investigators to question whether the endometrial cell apoptosis induced by GnRH analogues related to changes in the expression of pro- or anti-apoptotic proteins.

As mentioned previously, two major pathways towards apoptosis have been defined: the extrinsic death receptor pathway and the intrinsic mitochondrial pathway. The Bcl-2 family of proteins constitutes a critical intracellular checkpoint in the intrinsic pathway of apoptosis; some of these proteins, such as Bax, Bcl-xS, Bak, and Bok, have pro-apoptotic effects and others, such as Bcl-2, Bcl-xL, and Mcl-1, have anti-apoptotic effects. The ratio of pro- to anti-apoptotic molecules, such as Bcl-2/Bax, constitutes a rheostat that sets the threshold of susceptibility to apoptosis for the intrinsic pathway. Through this interaction, it has been proposed that Bax antagonizes Bcl-2 function, diminishing the ability of Bcl-2 to prolong cell survival.

To investigate this, the expression of Bax and Bcl-2 protein in EEC cultures from EDT patients were evaluated by Western blot and immunocytochemistry. Western blot analysis was performed on cell homogenates from EEC cell cultures from patients with EDT under basal conditions and after treatment with LA (1,000 ng/ml), ANT (10^{-5} M) or a combination of ANT with LA. The results of this investigation are shown in figure 8. In EEC from women with EDT, Bax expression increased after

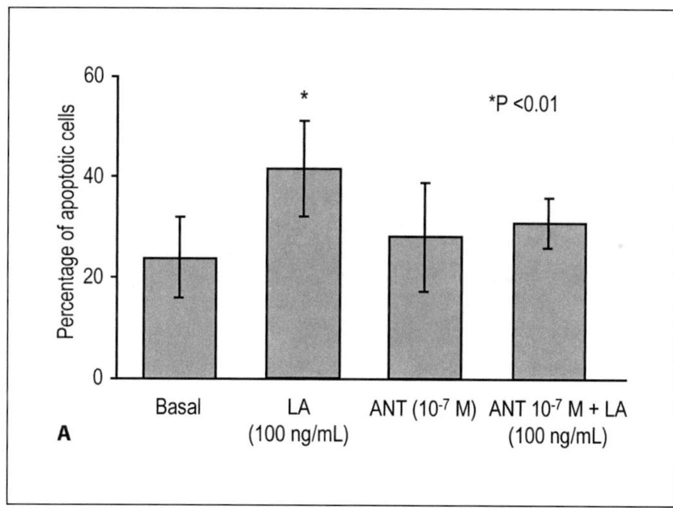
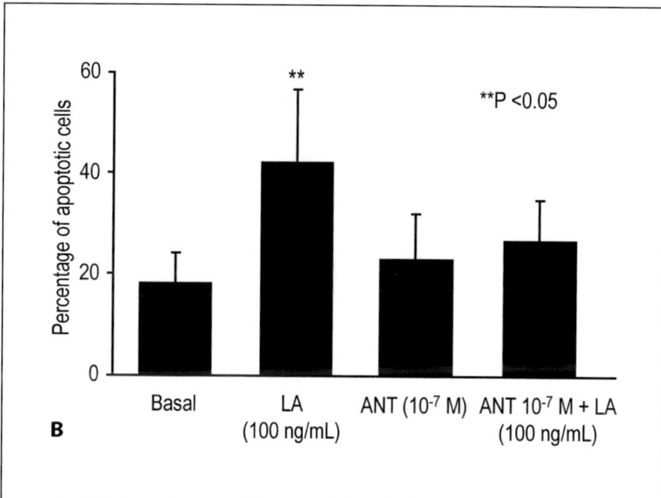

Fig. 6. Effect of leuprorelin acetate (LA) and antide (ANT) on the percentage of apoptotic cells in epithelial endometrial cell cultures obtained from EDT patients (**A**) or control subjects (**B**) [15]. Reproduced with permission from Meresman et al. [15].

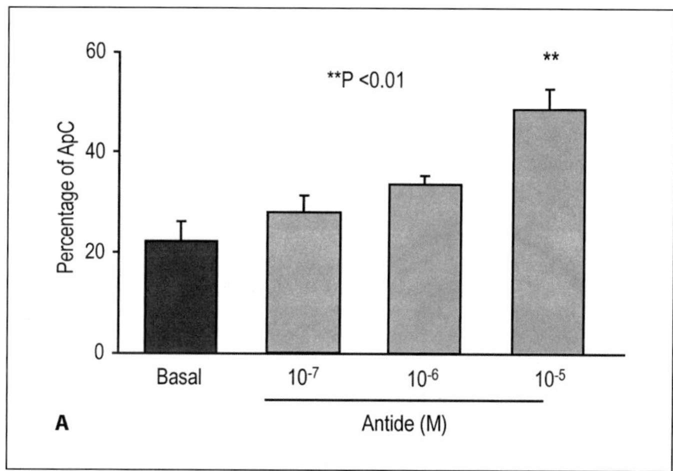
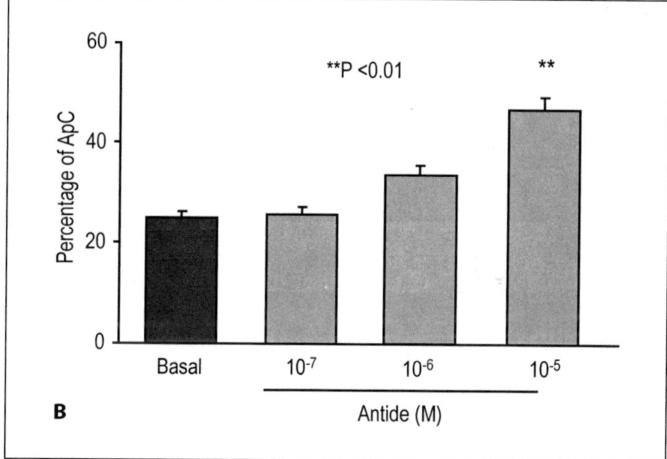

Fig. 7. Effects of antide on apoptosis in endometrial cell cultures from subjects with EDT (**A**) and controls (**B**). Values are expressed as percentage of apoptotic cells (ApC) ± SEM [14]. Reproduced with permission from Bilotas et al. [14].

treatment with LA, ANT and LA + ANT ($p < 0.05$, $p < 0.001$ and $p < 0.001$) (fig. 8A), whereas Bcl-2 expression decreased after exposure to LA and ANT ($p < 0.001$ and $p < 0.01$) (fig. 8B).

In relation to the extrinsic apoptosis pathway, the Fas–Fas ligand (FasL) system is the primary mechanism for induction of apoptosis in cells and tissues. Expression of Fas and FasL protein in EEC cultures from EDT patients were evaluated by Western blot and immunocytochem-istry. FasL expression increased after LA, ANT and LA + ANT treatments ($p < 0.01$, $p < 0.001$ and $p < 0.01$) (fig. 9A), but no significant changes were observed in Fas expression (fig. 9B).

From these results it can be concluded that GnRH analogues enhance apoptosis in EEC, and that this is accompanied by an increase in the expression of the pro-apoptotic proteins Bax and FasL and a decrease in expression of the anti-apoptotic protein Bcl-2.

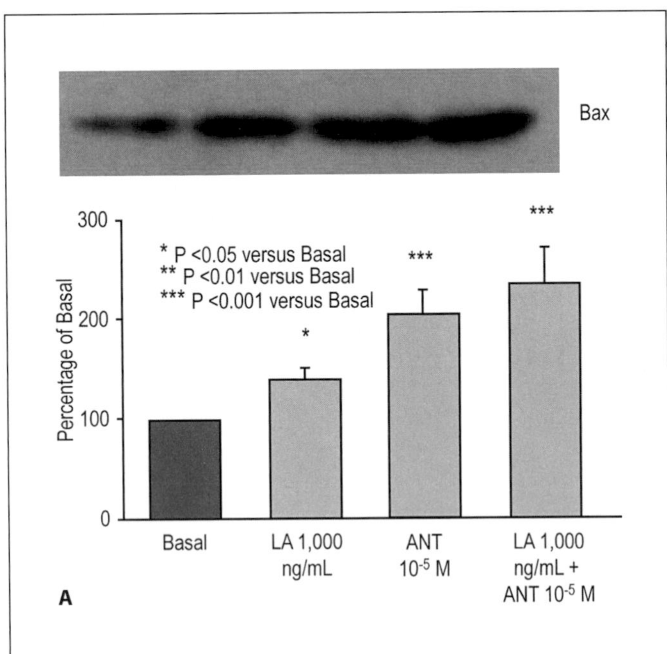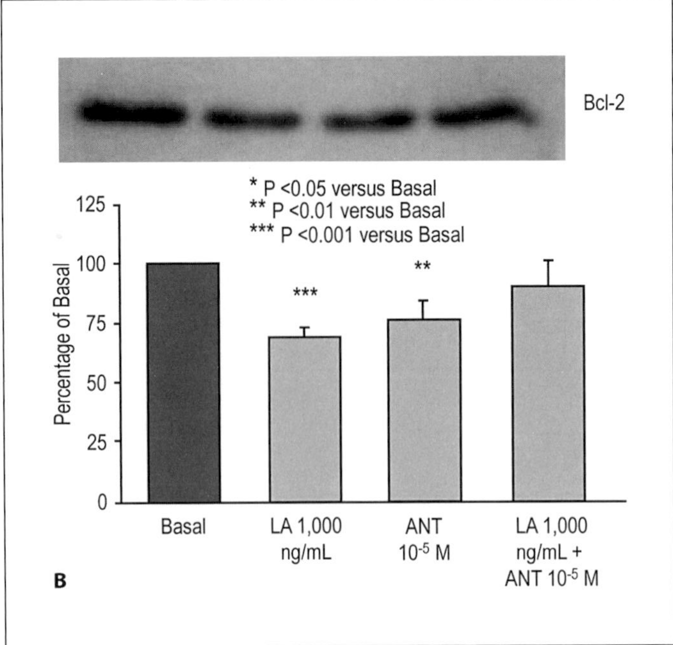

Fig. 8. Effects of leuprolide acetate (LA) and antide (ANT) on Bax (**A**) and Bcl-2 (**B**) protein expression in endometrial cell cultures obtained from EDT patients. Representative Western blots are presented in upper panels and quantification of results for protein expressed as a percentage of basal values $\pm$ SEM in the lower panels [14]. Reproduced with permission from Bilotas et al. [14].

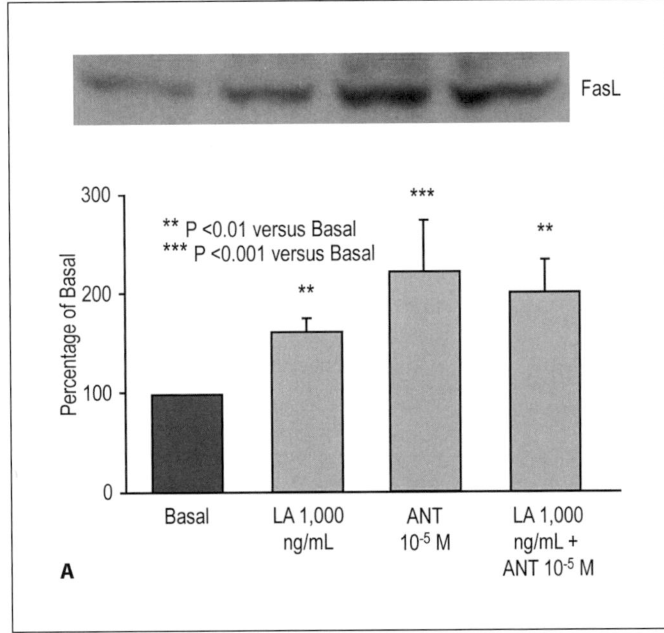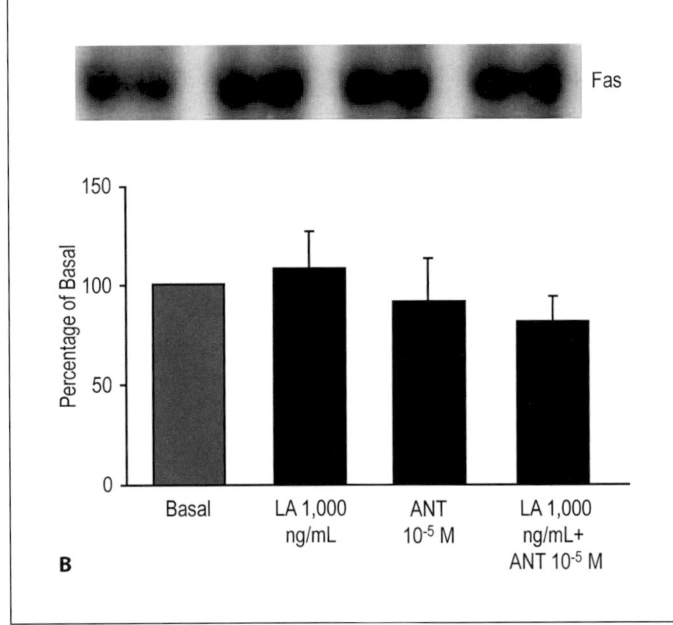

Fig. 9. Effects of leuprolide acetate (LA) and antide (ANT) on FasL (**A**) and Fas (**B**) protein expression in endometrial cell cultures obtained from EDT patients. Representative Western blots are presented in upper panels and quantification of results for protein expressed as a percentage of basal values $\pm$ SEM in the lower panels [14]. Reproduced with permission from Bilotas et al. [14].

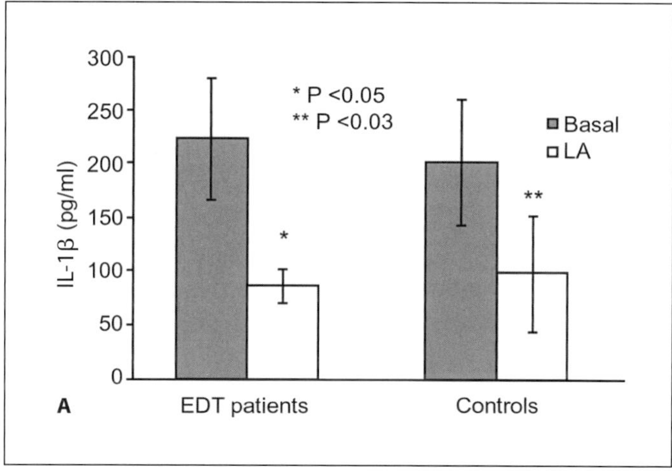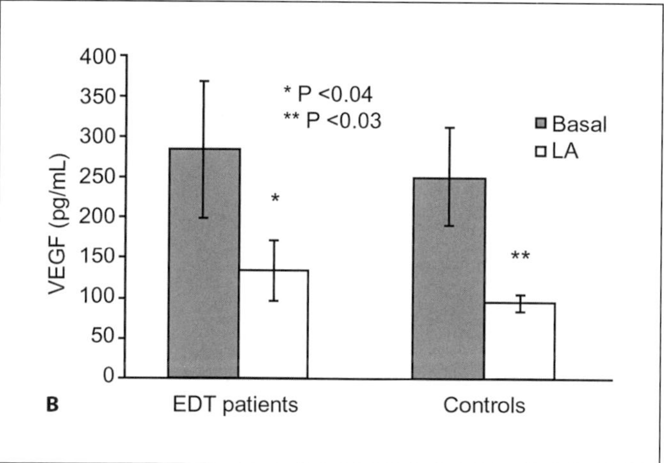

Fig. 10. Effects of leuprolide acetate (LA) on IL-1β (**A**) and VEGF (**B**) production by endometrial cell cultures obtained from EDT patients and control subjects [16]. Reproduced with permission from Meresman et al. [16].

Studies of GnRH Analogues and Angiogenesis

Studies have also been undertaken to evaluate the effect of GnRH analogues on the production of angiogenic factors in endometrial tissue [16]. It was questioned whether the production of VEGF and/or IL-1β is regulated by the presence of LA in the culture medium of endometrial cells.

Biopsy specimens of eutopic endometrium obtained from 16 women with untreated EDT and 14 controls were studied. IL-1β and VEGF release were evaluated in EEC cultures under basal conditions or with the addition of LA (10^{-7} M). IL-1β and VEGF concentrations were assessed by using commercial enzyme-linked immunosorbent assay (ELISA) kits. IL-1β and VEGF release were downregulated by LA in cultures from controls and EDT patients (fig. 10A, B).

The authors concluded that GnRH agonists appear to have a direct effect in endometrial cell cultures, in part by decreasing the release of pro-mitogenic cytokines such as IL-1β and VEGF.

Conclusions

From these preclinical studies of pathogenic mechanisms involved in EDT it can be concluded that LA inhibits cell proliferation and increases the apoptotic rate in EEC cultures. The GnRH antagonist, ANT, is able to reverse these effects, suggesting that the effect of GnRH analogues on DNA synthesis and apoptosis is mediated by homologous receptors. Higher ANT concentrations (10^{-5} M) enhance apoptosis in EEC cultures, most likely via another receptor distinct from the classic GnRH receptor. GnRH analogues also increase the expression of the pro-apoptotic proteins Bax and FasL and decrease the expression of the anti-apoptotic protein Bcl-2. LA inhibits the release of VEGF-A and IL-1β in eutopic endometrial cell cultures, factors known to be involved in angiogenesis. Thus, GnRH analogues are effective in reducing the growth of endometrial cells not only by the classical pituitary endocrine effect, but also via a direct effect on the endometrial cells themselves. Ultimately, this knowledge may contribute to the development of improved, targeted therapies for this debilitating disease.

Post-Presentation Discussion

Prof. Terakawa, Tottori University, Japan: Are there differences between GnRH agonists and antagonists in the induction of apoptosis?

Dr. Tesone: The effect is the same but in our experiments the GnRH antagonists, antide, only produced effects at higher concentrations (10^{-5} M). It is thought that the effect of antide is mediated through another receptor. When we add antide at lower concentrations, the effect of LA is blocked, suggesting that they are acting at the same receptor and LA is not able to bind in the presence of antide. At higher concentrations it appears that antide binds to another receptor, the identity of which is currently unknown.

Dr. Kajitani, Keio University, Japan: You showed the direct effect of GnRH analogues mediated via the GnRH receptor on endometrial cells. As a consequence of binding to the GnRH receptor, the protein kinase-C pathway will be activated; however, you did not show any direct effect on products of the protein kinase-C pathway. Have you any information about the protein kinase-C pathway in this context?

Dr. Tesone: That is a very interesting question. We have not studied the protein kinase-C pathway in this experimental model, but have done so in ovarian cells. We found a direct effect of GnRH analogues on ovarian cells and an activation of the protein kinase-C pathway.

Dr. Harada, Tottori University, Japan: Did you investigate the effects of other GnRH agonists, such as buserelin or nafarelin?

Dr. Tesone: We have only undertaken studies with LA as it is the agonist most frequently used in the clinical setting in Argentina for the treatment of different pathologies.

Dr. Harada, Tottori University, Japan: The reason I ask is that we undertook studies with buserelin and found no effect on endometrial cell proliferation, suggesting this effect could be specific to LA.

Dr. Kahn, Nagasaki University, Japan: You investigated the pro-apoptotic marker, Bax, and the anti-apoptotic marker, Bcl-2, in EEC cultures from both control women and EDT patients. Which biological effect is stronger? Did you measure the ratio of Bax/Bcl-2? Unless you know this ratio, it is difficult to understand the overall effect of the GnRH agonist.

Dr. Tesone: We did measure the ratio and confirmed that with LA there is an increase in pro-apoptotic proteins and a decrease in anti-apoptotic proteins.

Dr. Kitawaki, Kyoto University of Medicine, Japan: The molecular structure of antide is similar to LA. In view of this, you might anticipate that antide would have some effect of its own on apoptosis, in addition to antagonizing the effect of LA. It may be valuable to evaluate the effect of antide at lower doses.

Dr. Tesone: We have investigated lower doses of antide and it has no effect other than inhibiting the action of LA.

Disclosure Statement

Dr. Marta Tesone is an investigator and has received honoraria from Abbott. The other authors of this chapter (Dr. R.I. Barañao, Dr. G. Meresman and Dr. M. Bilotas) do not have any relationships to disclose.

References

1 Borroni R, Di Blasio AM, Gaffuri B, Santorsola R, Busacca M, Viganò P, Vignali M: Expression of GnRH receptor gene in human ectopic endometrial cells and inhibition of their proliferation by leuprolide acetate. Mol Cell Endocrinol 2000;159:37–43.

2 Tesone M, Vitale A, Parborell F: Gonadotropin-releasing hormone action on ovarian steroidogenesis and apoptosis. Research Trends 1999;2:81–91.

3 Emons G, Schally AV: The use of luteinizing hormone releasing hormone agonists and antagonists in gynaecological cancers. Hum Reprod 1994;9:1364–1379.

4 Grundker C, Gunthert AR, Westphalen S, Emons G: Biology of the gonadotropin-releasing hormone system in gynecological cancers. Eur J Endocrinol 2002;146:1–14.

5 Grundker C, Schlotawa L, Viereck V, Eicke N, Horst A, Kairies B, Emons G: Antiproliferative effects of the GnRH antagonist cetrorelix and of GnRH-II on human endometrial and ovarian cancer cells are not mediated through the GnRH type I receptor. Eur J Endocrinol 2004;151:141–149.

6 MacFarlane M, Williams AC: Apoptosis and disease: a life or death decision. EMBO Rep 2004;5:674–678.

7 Stouffer RL, Martínez-Chequer JC, Molskness TA, Xu F, Hazzard TM: Regulation and action of angiogenic factors in the primate ovary. Arch Med Res 2001;32:567–575.

8 Donnez J, Smoes P, Gillerot S, Casanas-Roux F, Nisolle M: Vascular endothelial growth factor (VEGF) in endometriosis. Hum Reprod 1998;13:1686–1690.

9 McLaren J: Vascular endothelial growth factor and endometriotic angiogenesis. Hum Reprod Update 2000;6:45–55.

10 Lebovic DI, Bentzien F, Chao VA, Garrett EN, Meng YG, Taylor RN: Induction of an angiogenic phenotype in endometriotic stromal cell cultures by interleukin-1beta. Mol Hum Reprod 2000;6:269–275.

11 Stocks J, Bradbury D, Corbett L, Pang L, Knox AJ: Cytokines upregulate vascular endothelial growth factor secretion by human airway smooth muscle cells: Role of endogenous prostanoids. FEBS Lett 2005;579:2551–2556.

12 Meresman GF, Vighi S, Buquet RA, Contreras-Ortiz O, Tesone M, Rumi LS: Apoptosis and expression of Bcl-2 and Bax in eutopic endometrium from women with endometriosis. Fertil Steril 2000;74:760–766.

13 Meresman GF, Augé L, Barañao RI, Lombardi E, Tesone M, Sueldo C: Oral contraceptives suppress cell proliferation and enhance apoptosis of eutopic endometrial tissue from patients with endometriosis. Fertil Steril 2002;77:1141–1147.

14 Bilotas M, Barañao RI, Buquet R, Sueldo C, Tesone M, Meresman G: Effect of GnRH analogues on apoptosis and expression of Bcl-2, Bax, Fas and FasL proteins in endometrial epithelial cell cultures from patients with endometriosis and controls. Hum Reprod 2007;22:644–653.

15 Meresman GF, Bilotas M, Buquet RA, Barañao RI, Sueldo C, Tesone M: Gonadotropin-releasing hormone agonist induces apoptosis and reduces cell proliferation in eutopic endometrial cultures from women with endometriosis. Fertil Steril 2003;80(suppl 2):702–707.

16 Meresman GF, Bilotas MA, Lombardi E, Tesone M, Sueldo C, Barañao RI: Effect of GnRH analogues on apoptosis and release of interleukin-1beta and vascular endothelial growth factor in endometrial cell cultures from patients with endometriosis. Hum Reprod 2003;18:1767–1771.

Gynecologic and
Obstetric Investigation

Gynecol Obstet Invest 2008;66(suppl 1):19–27
DOI: 10.1159/000148027

Published online: October 21, 2008

Current Approaches to Optimizing the Treatment of Endometriosis in Adolescents

Marc R. Laufer

Division of Gynecology, Children's Hospital Boston, Department of Obstetrics, Gynecology and Reproductive
Biology, Brigham and Women's Hospital, and Harvard Medical School, Boston, Mass., USA

Key Words
Endometriosis · Treatment · Adolescents

Abstract
Endometriosis can occur in adolescents and this patient group presents particular challenges in terms of differential diagnosis, variable presentation and symptoms, and choice of treatment. Early diagnosis is essential in order to decrease pain and hopefully prevent disease progression and preserve future fertility. Endometriosis surgery is generally cytoreductive rather than curative, and postoperative medical therapy should be initiated regardless of disease stage. Menstrual suppressive therapy with the use of continuous combination estrogen/progestin is the main treatment for most adolescents with endometriosis. For those with a persistence of pain on this therapy Gonadotropin-releasing hormone (GnRH) agonists (with add-back therapy) can be effective in relieving symptoms. GnRH agonist therapy requires special consideration in adolescents due to possible adverse effects on bone mineralization – an important consideration in adolescents who are at a critical age for accrual of bone mineral density (BMD). However, potential problems of bone loss may be avoided with the use of 'add back' therapy. A recent clinical study found that most adolescents with endometriosis receiving a GnRH agonist plus add-back therapy with norethindrone acetate (NA) or estrogen plus NA had normal BMD at the hip. Add-back therapy appears to be a promising adjunct to GnRH agonist therapy for the prevention of bone loss and may allow a longer duration of therapy than with a GnRH agonist alone. BMD should continue to be carefully monitored after the initial 6–8 month period of therapy and then approximately every two years in adolescent patients (over age 16) receiving long-term GnRH agonist with add-back therapy.
Copyright © 2008 S. Karger AG, Basel

Introduction

This paper will review the clinical aspects of endometriosis and issues specifically related to management of the disease in adolescents. Pain related to endometriosis is not a new concept and has been observed in patients for many years. It is now being increasingly recognized that endometriosis is not just a disease of adult women, and does occur in adolescents.

Endometriosis is an enigma and its diagnosis is challenging. The definitive diagnosis is often debated and its exact etiology is still unknown. The presentation of endometriosis in adolescents is variable, and can be different to that observed in adults. While there are a range of

Marc R. Laufer
Division of Gynecology, Children's Hospital Boston
Boston, MA 02115 (USA)
Tel. +1 617 355 5785, Fax +1 617 730 0186
E-Mail marc.laufer@childrens.harvard.edu

treatment options available for the management of endometriosis, the long-term outcomes of therapy have not been well studied in adults, and there are no specific published data for treatment of adolescents, although preliminary results from some unpublished studies are now available and will be reviewed here.

Differential Diagnosis of Endometriosis

Endometriosis is classically defined as the presence of endometrial tissue in an abnormal location. It is known that endometrial tissue has receptors for estrogen, progestin, and androgen, and that it is hormonally responsive tissue.

Various theories have been debated to explain the etiology of the disease, including Sampson's theory of retrograde menstruation, hematogenous and/or lymphatic spread of endometrial tissue, metaplastic transformation, an immune system abnormality in ectopic endometrium, environmental exposure (for example to dioxin), and genetic predisposition. It is likely that there is not one single explanation and all of these factors play a role in disease development and progression.

Differential diagnosis is challenging in both adults and adolescents and there can be an overlap of different diseases, including those of the gastrointestinal system and urinary tract, pelvic inflammatory disease, ovarian cysts or tumors, and obstructive anomalies.

Similarly, the presenting symptoms can also be varied, ranging from no symptoms at all to dysmenorrhea, localized pain, dyspareunia and noncyclic pain. Adult women may also present with subfertility or infertility (seen at stage I to stage IV disease) or with gastrointestinal (constipation, diarrhea, hematochezia, melena) or genitourinary (hematuria, dysuria, frequency, urgency) symptoms. Pelvic masses/cysts from enedometriosis (endometriomas) can be observed, but are not typically seen in patients younger than the age of 22 years.

In women who present with pelvic pain, reports have shown that endometriosis is observed in 72–82% of adults [1–4] and 69–74% of adolescents [5, 6]. It should be noted that these studies were undertaken in the early 1990s and since that time the visual definition of laparoscopic techniques has improved considerably, so these numbers are probably low. More recently, the International Endometriosis Association has reported results from a registry of 4,000 women with endometriosis. They found that most women see approximately nine different health care providers before being given a definitive diagnosis of endometriosis [7]. In addition, most women have had symptoms that they associate with endometriosis since adolescence: 21% had severe pain under the age of 15 years, 17% between the ages of 15 and19, and 50% at age 24 years or younger. These data suggest endometriosis is a chronic, progressive disease and that early diagnosis is an opportunity to intervene in this process.

Adolescent endometriosis is most commonly diagnosed in young postmenarcheal adolescents, but it has also been reported in girls who are premenarcheal. One study has reported cases of premenarcheal endometriosis occurring without an associated obstructive anomaly [8]. Three premenarcheal subjects (aged 8.5, 11 and 11 years) who had chronic (>6 months) pelvic pain but negative gastrointestinal findings (evaluation by a pediatrician and gastroenterologist, and a normal pelvic ultrasound scan) underwent laparoscopic biopsies and resection. Clear, red, and/or white lesions were identified consistent with endometriosis. Biopsies did not reveal glands or stroma and the patients' pain improved with surgical management. These findings support the concept that some cases of the disease may result from an etiology other than retrograde menses.

Evaluation of Pain

Physical examination can be challenging in adolescents and thus should not be mandated for an evaluation of pelvic pain; adolescents with pelvic pain may be apprehensive about speaking about their condition as they fear an internal examination. In view of this, it is important to be clear about what we need to achieve with any physical examination in these young women. The goal should be to rule out an obstructive anomaly or ovarian mass, but the usual bimanual examination may not be possible. One option in an adolescent is to undertake a Q-tip examination – inserting a Q-tip into the vaginal vault and ensuring that it goes to an adequate length, to rule out a vaginal obstructive anomaly such as a transverse vaginal septum. Pelvic ultrasound can be used to help rule out the possibility of an ovarian mass or congenital anomaly of the reproductive tract.

Characteristics of Endometriosis in Adolescents

A retrospective study has been undertaken in adolescent girls (<22 years of age) with chronic pelvic pain not responding to conventional medical therapy with oral contraceptives and nonsteroidal anti-inflammatory drugs (NSAIDs) [6]. These subjects underwent a laparoscopy to determine the etiology of their pain following referral to a surgical gynecologist in a pediatric/adolescent gynecology academic practice. More than two-thirds of the study population (69.6%) was found to have endometriosis, and all of these subjects had either stage I or II disease. The nature of the pain in the 32 subjects with endometriosis was both acyclic and cyclic in 20 (62.5%), acyclic only in 9 (28.1%), and cyclic only in 3 (9.4%). This is in contrast to pain generally observed in older women with endometriosis which tends to be cyclic in nature. Other presenting symptoms included gastrointestinal in 11 (34.3%), urinary in 4 (12.5%), and irregular menses in 3 (9.4%). The authors concluded that adolescents with chronic pelvic pain not responding to medical therapy have a high rate of endometriosis and should therefore be referred to a specialist gynecologist who can confirm a definitive diagnosis and initiate appropriate therapy.

Establishing a Diagnosis in Adolescents with Endometriosis

In view of these findings, what should our approach be to adolescents with chronic pelvic pain? In the first instance it is important that the adolescents themselves identify pain, that we initiate treatment of the dysmenorrhea and, if pain persists, establish a diagnosis of endometriosis as soon as possible, so the women do not suffer needlessly.

Endometriosis treatments can comprise both medical and surgical interventions, including the use of complementary and alternative medicines, and, importantly, pain management. The approach should be multidisciplinary and include emotional support for the young women and the provision of educational materials to help them better understand their condition and its treatment. Teen peer support programs and internet chat rooms have been found to be helpful for this age group (www.youngwomenshealth.org).

In order to provide guidance to the physician in the appropriate management of this patient group it would be helpful to develop algorithms for optimum diagnosis and treatment. Adolescents with dysmenorrhea should initially be treated with combination cyclic hormonal therapy and NSAIDs. If pain persists an evaluation for endometriosis should be initiated. Typically, the classical gold standard for diagnosis of endometriosis has been by sur-

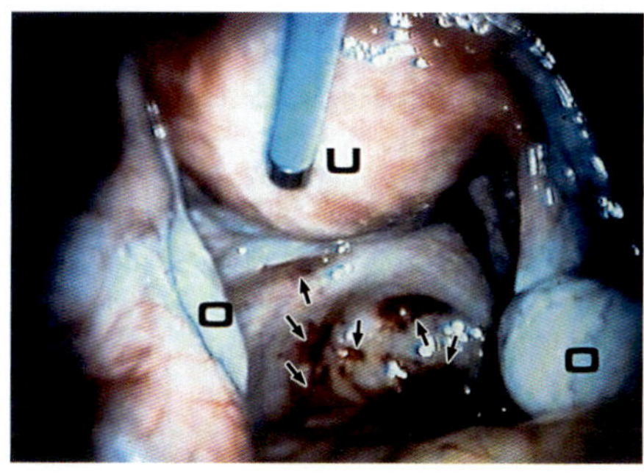

Fig. 1. Endometriosis in adolescents is commonly observed in the posterior cul-de-sac and on the utero-sacral ligaments. Reproduced with permission from Emans et al. [31].

gery, either visualization by laparoscopy or by biopsy [9]. Empiric use of GnRH agonist therapy is offered to adults and adolescents over age 18 [10], but GnRH agonist empiric therapy is not recommended for adolescents less than 18 years of age.

When managing adolescent patients with endometriosis, it is important to remember – and to reassure the patient – that no correlation has been found between severity of pain symptoms and stage of the disease or site of the endometriotic lesions [11]. Endometriosis can also occur in a range of different locations – the bowel and the bladder, for example – but the most common location in adolescents is in the posterior cul-de-sac (fig. 1). The black lesions or powder-burn lesions seen in adults are not commonly observed in adolescents, so from the absence of these we cannot infer that the problem is not endometriosis. Clear or red lesions are the most common lesions observed in adolescents and also found to be the most painful lesions [12].

Laparoscopy is the primary surgical approach for the management of endometriosis, and this can be either resection or ablation/destruction of the lesions – at stage I or II disease there is really little difference in these approaches – using a laser, scissors, harmonic scalpel, or electrosurgery. Hysterectomy is sometimes recommended by frustrated gynecologists due to the intractable nature of this disease; however, this approach is *not* appropriate for an adolescent population.

When diagnosing endometriosis it is important to recognize the wide range of types and colors of lesions that can be observed in this disease [13]. Clear lesions are common in adolescent endometriosis but often difficult to visualize and evaluate. Improved visualization and identification can often be achieved simply by increasing

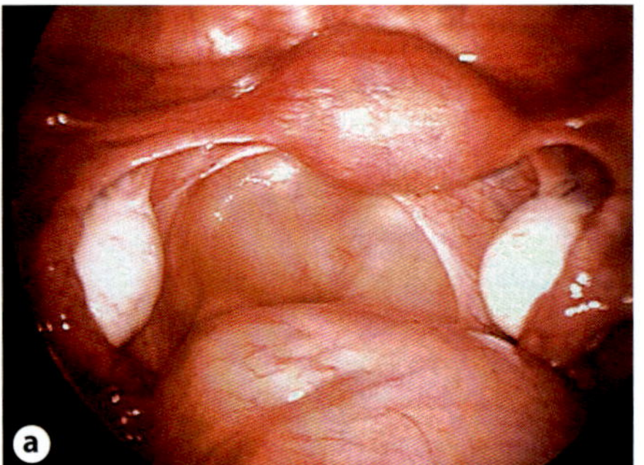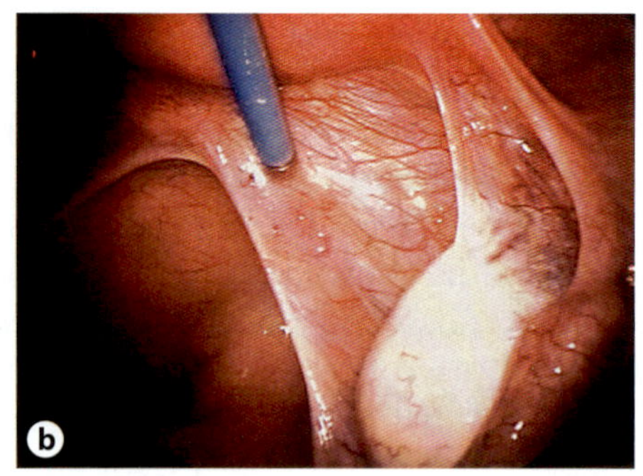

Fig. 2. Examples of clear and opaque endometriotic lesions. Magnification of clear endometriotic lesions improves visualization and identification: **a** apparently normal pelvis in a 15-year-old girl and **b** magnification to reveal clear lesions. **a** Reproduced with permission from the personal collection of Marc R. Laufer, M.D. **b** Reproduced with permission from Emans et al. [31].

Outcomes of Surgical Treatment

In adolescents there are no published data on the outcomes of surgical therapy for endometriosis, so data on outcomes in adults will be reviewed here. Reports of pain recurrence in adults following surgical therapy are high, ranging from 30–51% [15–19].

Adult studies of aggressive surgical therapy with laser ablation and laparoscopic uterosacral nerve ablation (LUNA) compared with diagnostic laparoscopy alone have shown that after one year 56% of patients treated show improvement while the remainder still have pain (fig. 3) [19, 20]. These studies also show that stage I disease shows the least improvement with surgery (38% of patients after 6 months) while stage III is most improved (100% of patients after 6 months) [19, 20].

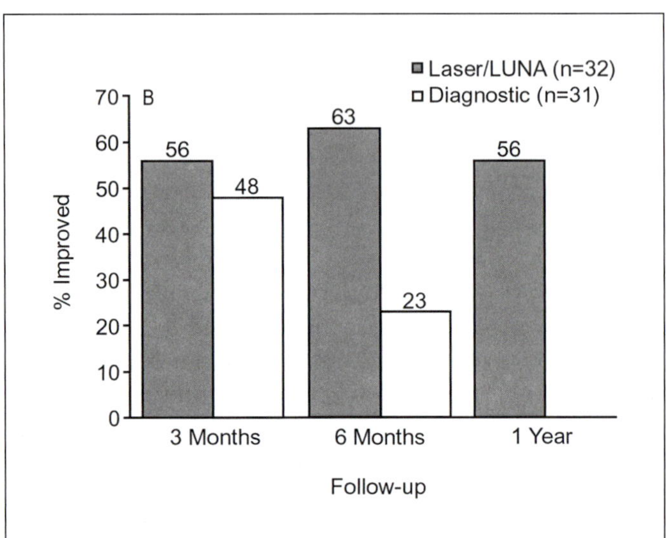

Fig. 3. Improvement rates in endometriotic pain following aggressive surgical therapy with laser ablation and laparoscopic uterosacral nerve ablation (LUNA) compared with diagnostic laparoscopy alone [19, 20].

the laparoscopic magnification. Figure 2 shows an example of pelvic laparoscopy in a 15-year-old girl. The pelvis appears normal (a), but if magnification is increased, clear endometriotic lesions become visible (b). A new technique has now been described to improve the visualization and identification of clear vesicles of endometriosis [14]. The pelvis is filled with irrigation fluid, and then the laparoscope is submerged, with evaluation of the pelvic peritoneum through the liquid medium. This technique enables the viewer to appreciate the three-dimensional configuration of the lesions and to exclude light reflection as a possible etiology. Peritoneal defects, or windows, which are also diagnostic of endometriosis, are very common in adolescents.

Medical Therapy for Adolescents with Endometriosis

A range of medical therapies have been employed for the treatment of endometriosis, although data are not available for all of these options in an adolescent population. The options include estrogen/progestin combinations [continuous administration of oral contraceptives via pills (OCPs), or vaginal rings], progestins alone (Depo-Provera, Aygestin), and GnRH agonists with add-back therapy [commonly Aygestin (5 mg) or PremPro (0.625 mg/2.5 mg)]. It is important to note that when add-back therapy is used with a GnRH analogue, OCPs should not be used.

The mechanism of action of OCPs in the treatment of endometriosis is to induce ovulation inhibition, decidualization of the endometrium, and decreased menstrual flow. This therapy is the usual first-line treatment for adolescent endometriosis and can be very successful. When administering continuous combined OCPs to an adolescent patient it is important to explain in detail how to take the pills – continuously rather than cyclically, as they would for contraception.

In the case of progestins only, studies with medroxyprogesterone acetate (MPA), norethindrone acetate, megestrol and gestrinone have reported symptom improvement in around 80% of patients and disease improvement in around 60% [21–23]. If using an injectable formulation of progestin, it is important to consider possible effects on bone density, especially in the adolescent population, and the need for estrogen add-back therapy in case of long-term therapy. However, studies with MPA suggest that its ability to induce amenorrhea is relatively poor (70% of patients) and patients often complain of side effects: weight gain (60%), bloating and edema (60%), anxiety or irritability (20%) and dysfunctional bleeding (20%) [21]. The Endometriosis Association data reported that this was not a very successful treatment option for adult women [7]. In another study, ratings of endometriosis medical therapies by adult patients show that progestins are not considered very effective, whereas pain medications followed by GnRH agonists are most effective [7].

Danazol is effective in treating the pain associated with endometriosis and in suppressing menstruation, but in an adolescent population it is not an acceptable choice of therapy due to its androgenic side effects, such as weight gain or deepening of the voice [24, 25].

The GnRH agonists (buserelin, goserelin, histrelin, leuprorelin and naferelin) have been found to be highly effective in the management of endometriosis with reductions of pain in 80% of adult patients treated with leuprorelin [26]. The side effects of GnRH agonists include vasomotor symptoms, such as hot flushes, reductions in bone mineral density (BMD), mood swings and vaginal dryness. These side effects can be alleviated if clinicians routinely prescribe add-back therapy with the initiation of the GnRH agonist therapy. When prescribing an agonist, it is important that the doctor explains to the patient exactly how the medication works and the flare effect that can be experienced. Ideally, the GnRH agonist should be initiated in the late luteal phase of the menstrual cycle. If switching from OCPs, there should be an overlap of about three weeks with the GnRH agonist in an attempt to suppress the flare effect. OCPs should not be used as add-back therapy with GnRH agonists.

As illustrated in figure 4, if add-back therapy is employed and the correct sex steroid concentrations are achieved – the therapeutic window – then bone loss will

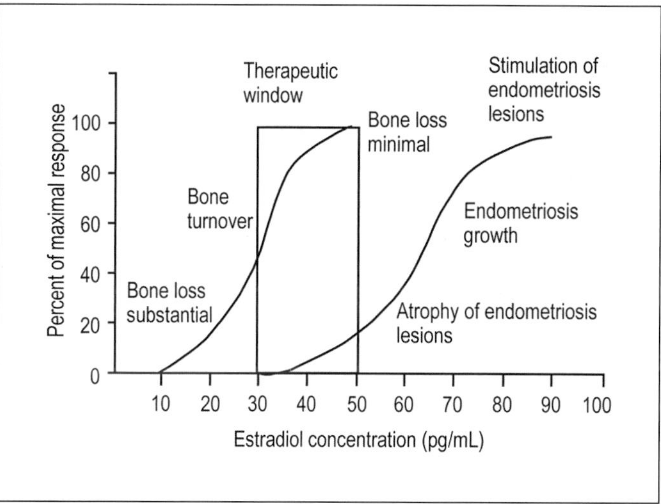

Fig. 4. 'Therapeutic window' for add-back therapy in order to minimize both bone loss and proliferation of endometriosis lesion. Adapted from Barbieri et al. [27].

be minimized, but endometrial growth will not be stimulated [27].

There are some important considerations with the use of GnRH agonists in adolescents with endometriosis. Adolescence is a critical time period for bone mineral accrual and there are many issues that affect adolescent bone health, including endocrine factors, diet, eating disorders, and exercise. All adolescents on GnRH agonists should take calcium and vitamin D supplementation in addition to the recommended add-back therapy. When testing BMD in adolescents, it is important to use Z-scores, rather than the T-scores used in adults. T-scores compare BMD to peak bone mass at 20 years [28]. For an adolescent the T-score will be low compared to a normal mean BMD of a 20 year old and thus be inappropriately diagnosed as osteoporosis. In contrast, Z-scores compare with age-matched peers and are thus more accurate for measurement in adolescents. If GnRH therapy is planned for 8 months or longer, BMD should be checked, typically every two years.

In the light of the concerns about the effects of GnRH agonists on BMD, a study was undertaken to assess the bone density of adolescent and young adult women with endometriosis treated with a GnRH-agonist and add-back therapy with norethindrone acetate [29]. In this retrospective study the charts of all patients with endometriosis referred to the Bone Health Center by the Gynecology Program at the Children's Hospital Boston, USA, between 1995–2005 were analyzed (n = 50). A total of 36

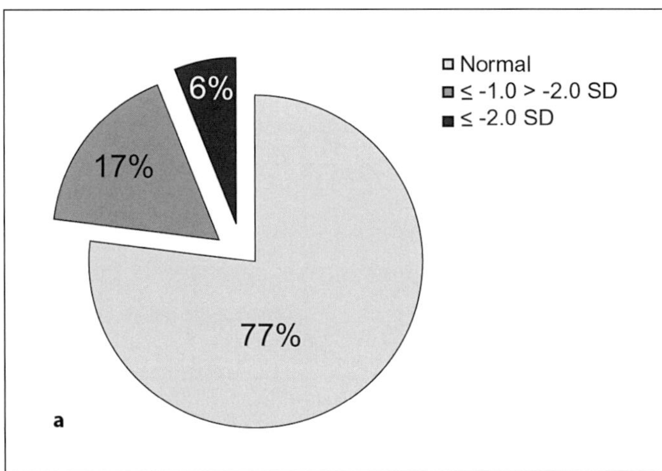
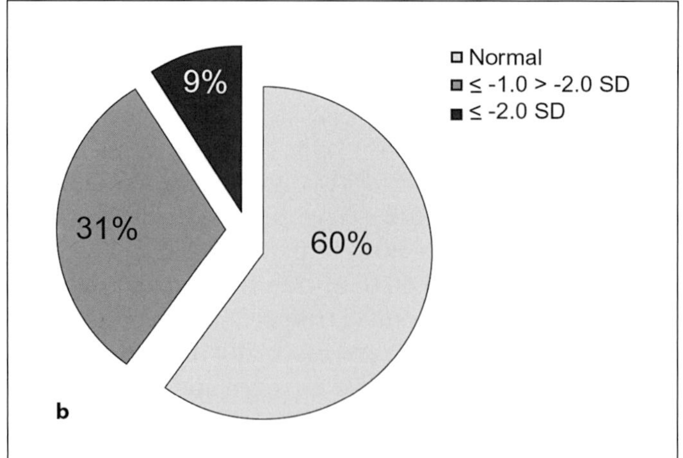

Fig. 5. BMD at the hip (**a**) and lumbar spine (**b**) in patients with a surgical diagnosis of endometriosis undergoing therapy with leuprolide acetate (3-month depot) and adjunctive treatment with norethindrone acetate (5 mg daily). Reproduced with permission from Divasta et al. [29].

Table 1. Characteristics of patients with a surgical diagnosis of endometriosis undergoing therapy with leuprolide acetate (3-month depot) and adjunctive treatment with norethindrone acetate (5 mg daily) at the time of dual energy X-ray absorptiometry (DXA) scanning

	Mean ± SD	Range
Age, years	17.7 ± 1.9	13.0–21.3
Weight, kg	62.9 ± 12.9	38.9–95.2
Height, cm	160.5 ± 6.4	143.4–177.8
BMI	24.5 ± 4.9	15.0–37.7
Duration of treatment, days	392 ± 276	140–1,184

Reproduced with permission from Divasta et al. [29].

patients with a surgical diagnosis of endometriosis undergoing therapy with leuprolide acetate (3-month depot) and adjunctive treatment with norethindrone acetate (5 mg daily) were included in the study. Dual energy X-ray absorptiometry (DXA) was used to measure total hip and lumbar spine BMD. The patients' height and weight were measured in standard fashion at the time of DXA scanning. Z-scores [±standard deviations (SD) from the mean] were calculated and interpreted as follows: –1.0 SD: normal range; ≤–1.0 to >–2.0: at risk for low bone mass; ≤–2.0 SD: significant low bone mass. Table 1 shows the patients' characteristics at the time of DXA scanning. Results for BMD at the hip and lumbar spine are shown in figure 5a and b.

The results showed no correlation between duration of therapy with the GnRH-agonist plus add-back and BMD at the hip or spine. BMD at the hip was normal in most adolescents with endometriosis who were receiving the GnRH agonist plus add-back therapy with norethindrone acetate. However, almost one-third of subjects in the study exhibited skeletal deficits at the spine and the authors concluded that BMD should be carefully monitored in adolescents receiving treatment with GnRH agonists. Add-back therapy is therefore an intriguing option for the preservation of skeletal health in adolescent patients being treated for endometriosis with GnRH agonists, but further prospective studies are needed to determine its efficacy.

To avoid any potential problems with the use of GnRH agonists in adolescents, it is important to educate them about the possible side effects. The use of a 3-month depot formulation may avoid them stopping therapy if they dislike the flare side effects. It should also be noted that most of the negative information about the use of GnRH agonists that patients might read on the internet, for example, may relate to the use of GnRH agonists alone and not GnRH agonists plus add-back therapy.

If bleeding occurs during GnRH agonist therapy after the initial withdrawal bleed there may be inadequate suppression. An estradiol level is helpful in determining if there has been adequate suppression. If there is not adequate suppression the dose of GnRH agonist may need to be increased. It may also be reassuring for patients and their parents that if they do not like the effects they experience with the 3-month depot formulation of the GnRH it is possible to add an OCP at any time and thus have the same side effect profile as the OCP alone.

In terms of outcomes of medical therapy in adolescents with endometriosis, there is only one published study to date [30]. Doyle and colleagues investigated 90 eligible patients aged 12–24 years who had undergone laparoscopy for diagnosis and surgical management of their condition. They received medical treatment with continuous combined estrogen/progestin (91% of subjects), leuprolide acetate plus norethindrone acetate or PremPro (78%), or progestin only (12%), many receiving multiple therapies depending on their symptoms. Patients included in the study then had a second laparoscopy due to a recurrence of pain. Results showed that a trend towards disease progression was not seen in this patient group and the authors concluded that standard surgical–medical management retards disease progression in young women with endometriosis.

Complementary and alternative therapies can also play a role in the management of adolescent endometriosis, and these approaches include herbal medicines and acupuncture. In addition, the beneficial effects of support networks for young women suffering from this disease should not be underestimated. A range of education and support materials is available from the Endometriosis Association (www.endometriosisassn.org) and a specific website focused on young women's health, including endometriosis, has been developed by the Children's Hospital Boston, USA (www.youngwomenshealth.org). Various written materials are available to download and users can also participate in chat rooms with other young women going through the same experience.

Conclusions

Endometriosis, while generally considered as a disease that affects adult women, has become increasingly recognized as a chronic illness that can begin during adolescence and young adulthood. In this patient group, early diagnosis is the key to achieving decreased pain and limiting the progression of the disease and preserve future fertility. The surgical option should be seen as cytoreductive rather than curative, and postoperative medical therapy should be initiated regardless of the disease stage. Ideally, there should be an attempt to limit life-time surgical interventions in these women. Despite the considerable advances in our understanding of endometriosis in adolescents, there are still several unanswered questions that need to be addressed in future research. These include whether early diagnosis and treatment preserves future fertility, and how we should best manage young women whose mothers have a history of infertility, stage IV endometriosis, but no pelvic pain.

Post-Presentation Discussion

Dr. Osuga, University of Tokyo, Japan: In terms of checking BMD in women receiving GnRH agonists plus add-back therapy you mentioned that for a duration of therapy of up to 8 months it may not be necessary, but for longer duration checking every two years was advisable. Does this also apply to adolescent patients? In adolescents the bone mass is increasing, so is a 2-year interval sufficiently frequent?

Dr. Laufer: We do not know the definitive answer to that question. The ACOG guidelines on adolescent endometriosis do not recommend the use of empiric GnRH agonist therapy in women under the age of 16 years. In my hospital we do use GnRH agonist therapy following a diagnosis of endometriosis over the age of 16 if they have a persistence of pain on combination continuous estrogen/progestin therapy and also in some younger patients who have failed on other therapies. There is currently no recommended interval for BMD measurements in these patients. We currently check a BMD if a patient elects to stay on GnRH agonist with add-back for longer than 8 months and if it is normal and they elect to continue the GnRH agonist with add-back we repeat the BMD every 2 years. We also recommend calcium and vitamin D supplementation while on GnRH agonist with add-back. Depending on the patients' individual circumstances and family history you may decide to measure more frequently than every two years; if we notice a decrease in BMD we might decide to check more often. This should be judged on a case by case basis since it is outside of recommended guidelines.

Dr. Wang, Taipei, Taiwan: I agree that we should treat adolescents with endometriosis with GnRH agonists plus add-back therapy. Based on results from the Women's Health Initiative (WHI) in 2002, hormone therapy is not recommended in menopausal women if they are asymptomatic. Why would we prescribe add-back therapy in adolescents receiving GnRH therapy when the WHI do not recommend it for postmenopausal women?

Dr. Laufer: It is important to consider the level of hormone that we are administering in adolescent patients. In postmenopausal women the studies have shown it is not beneficial to continue exposure to hormones after the normal menopausal stage. In adolescents the hormonal levels in those receiving GnRH agonist with add-back therapy is much lower that they would normally be exposed to during a natural cycle and there is no contraindication to using hormones at that level at that age. In both adults and adolescents I always use add-back therapy when I treat them with a GnRH agonist. There is no reason that a woman should suffer the hypoestrogenic side effects and the risk of bone loss if she is premenopausal.

Dr. Kitawaki, Kyoto University of Medicine, Japan: The ACOG guidelines mention that GnRH agonists should not be used empirically in patients aged under 16 years of age. My question is: how frequently do you treat empirically with GnRH agonists plus add-back therapy as opposed to following a diagnostic confirmation by laparoscopy?

Dr. Laufer: The recommendation you refer to in ACOG guidelines was not based on definitive data at that time. As an author of the guidelines, when I was asked the specific question: when would I feel comfortable treating an adolescent with laparoscopically diagnosed endometriosis with GnRH agonists plus add-back, I consulted my pediatric and endocrinology specialists and we decided on 16 years of age, but we do not have clinical data to support this yet. The ACOG guidelines specify empiric treatment with GnRH agonists without laparoscopic diagnosis only in patients over the age of 18 years from a medicolegal standpoint, i.e. when these patients are legally able to make their own decisions.

Prof. Hoshiai, Kinki University, Japan: Are there any other early diagnostic methods, other than laparoscopy, that can be employed in adolescents, for example Ca^{125}?

Dr. Laufer: Ca^{125} is very nonspecific and to my knowledge there is currently no serum marker that will detect endometriosis with a satisfactory level of accuracy and specificity. I always make a diagnosis of adolescent endometriosis with laparoscopy and I laparoscope all my patients. Some could argue that if so many cases of stage I and II disease are missed on laparoscopy, there is little point in doing it and patients should all be treated medically. However, many parents want a confirmed diagnosis before their daughter starts medical therapy, which is one of the issues of using empiric therapy.

Disclosure Statement

Dr. Laufer is an advisory board member for Merck & Co., Inc. He has participated in the speakers bureau at Merck & Co., Inc. and TAP, Inc. As an investigator, Dr. Laufer has received grant and research support from Johnson & Johnson.

References

1 Martin DC, Hubert GD, Vander Zwaag R, el-Zeky FA: Laparoscopic appearances of peritoneal endometriosis. Fertil Steril 1989;51: 63–67.

2 Cornillie FJ, Oosterlynck D, Lauweryns JM, Koninckx PR: Deeply infiltrating pelvic endometriosis: histology and clinical significance. Fertil Steril 1990;53:978–983.

3 Koninckx PR, Meuleman C, Demeyere S, Lesaffre E, Cornillie FJ: Suggestive evidence that pelvic endometriosis is a progressive disease, whereas deeply infiltrating endometriosis is associated with pelvic pain. Fertil Steril 1991;55:759–765.

4 Koninckx PR, Martin DC: Deep endometriosis: a consequence of infiltration or retraction or possibly adenomyosis externa? Fertil Steril 1992;58:924–928.

5 Reese KA, Reddy S, Rock JA: Endometriosis in an adolescent population: the Emory experience. J Pediatr Adolesc Gynecol 1996;9: 125–128.

6 Laufer MR, Goitein L, Bush M, Cramer DW, Emans SJ: Prevalence of endometriosis in adolescent girls with chronic pelvic pain not responding to conventional therapy. J Pediatr Adolesc Gynecol 1997;10:199–202.

7 Ballweg ML: Big picture of endometriosis helps provide guidance on approach to teens: comparative historical data show endo starting younger, is more severe. J Pediatr Adolesc Gynecol 2003;16(suppl 3):S21–S26.

8 Marsh EE, Laufer MR: Endometriosis in premenarcheal girls who do not have an associated obstructive anomaly. Fertil Steril 2005; 83:758–760.

9 American College of Obstetricians and Gynecologists: Endometriosis in adolescents. ACOG Committee Opinion No 310. Obstet Gynecol 2005;105:921–927.

10 Ling FW: Randomized controlled trial of depot leuprolide in patients with chronic pelvic pain and clinically suspected endometriosis. Obstet Gynecol 1999;93:51–58.

11 Fedele L, Parazzini F, Bianchi S, Arcaini L, Candiani GB: Stage and localization of pelvic endometriosis and pain. Fertil Steril 1990;53:155–158.

12 Demco L: Mapping the source and character of pain due to endometriosis by patient-assisted laparoscopy. J Am Assoc Gynecol Laparosc 1998;5:241–245.

13 American Society for Reproductive Medicine: Revised American Society for Reproductive Medicine classification of endometriosis 1996. Fertil Steril 1997;67:817–821.

14 Laufer MR: Identification of clear vesicular lesions of atypical endometriosis: a new technique. Fertil Steril 1997;68:739–740.

15 Sutton C, Hill D: Laser laparoscopy in the treatment of endometriosis. A 5-year study. Br J Obstet Gynaecol 1990;97:181–185.

16 Redwine DB: Conservative laparoscopic excision of endometriosis by sharp dissection: life table analysis of reoperation and persistent or recurrent disease. Fertil Steril 1991; 56:628–634.

17 Hornstein MD, Gleason RE, Orav J, Haas ST, Friedman AJ, Rein MS, Hill JA, Barbieri RL: The reproducibility of the revised American Fertility Society classification of endometriosis. Fertil Steril 1993;59:1015–1021.

18 Howard FM: The role of laparoscopy in chronic pelvic pain: promise and pitfalls. Obstet Gynecol Surv 1993;48:357–387.

19 Sutton CJ, Ewen SP, Whitelaw N, Haines P: Prospective, randomized, double-blind, controlled trial of laser laparoscopy in the treatment of pelvic pain associated with minimal, mild, and moderate endometriosis. Fertil Steril 1994;62:696–700.

20 Sutton CJ, Pooley AS, Ewen SP, Haines P: Follow-up report on a randomized controlled trial of laser laparoscopy in the treatment of pelvic pain associated with minimal to moderate endometriosis. Fertil Steril 1997;68:1070–1074.

21 Luciano AA, Turksoy RN, Carleo J: Evaluation of oral medroxyprogesterone acetate in the treatment of endometriosis. Obstet Gynecol 1988;72:323–327.

22 Schlaff WD, Dugoff L, Damewood MD, Rock JA: Megestrol acetate for treatment of endometriosis. Obstet Gynecol 1990;75:646–648.

23 Hornstein MD, Gleason RE, Barbieri RL: A randomized double-blind prospective trial of two doses of gestrinone in the treatment of endometriosis. Fertil Steril 1990;53:237–241.

24 Buttram VC Jr, Belue JB, Reiter R: Interim report a study of danazol for the treatment of endometriosis. Fertil Steril 1982;37:478–483.

25 Henzl MR, Corson SL, Moghissi K, Buttram VC, Berqvist C, Jacobson J: Administration of nasal nafarelin as compared with oral danazol for endometriosis. A multicenter double-blind comparative clinical trial. N Engl J Med 1988;318:485–489.

26 Dlugi AM, Miller JD, Knittle J: Lupron depot (leuprolide acetate for depot suspension) in the treatment of endometriosis: a randomized, placebo-controlled, double-blind study. Lupron Study Group. Fertil Steril 1990;54:419–427.

27 Barbieri RL: Hormone treatment of endometriosis: the estrogen threshold hypothesis. Am J Obstet Gynecol 1992;166:740–745.

28 Gafni RI, Baron J: Overdiagnosis of osteoporosis in children due to misinterpretation of dual-energy x-ray absorptiometry (DEXA). J Pediatr 2004;144:253–257.

29 Divasta AD, Laufer MR, Gordon CM: Bone density in adolescents treated with a GnRH agonist and add-back therapy for endometriosis. J Pediatr Adolesc Gynecol 2007;20:293–297.

30 Doyle JO, Missmer SA, Laufer MR: The effect of standard surgical-medical intervention on the progression of endometriosis in an adolescent population. Presented at the NASPAG Annual Clinical Meeting, CA, 2008.

31 Emans SJ, Laufer MR, Goldstein DP: Pediatric and Adolescent Gynecology, 5th ed. Philadelphia, Lippincott, Williams and Wilkins Publishing Company, 2005.

Gynecologic and
Obstetric Investigation

Subject Index Vol. 66, Suppl. 1, 2008

KARGER

© 2008 S. Karger AG, Basel

Fax +41 61 306 12 34
E-Mail karger@karger.ch
www.karger.com

Accessible online at:
www.karger.com/goi